Iron Oxide Nanoparticles

Edited by Xiao-Lan Huang

Published in London, United Kingdom

Iron Oxide Nanoparticles
http://dx.doi.org/10.5772/intechopen.95129
Edited by Xiao-Lan Huang

Contributors
Ankur Soam, Rajan Lakra, Rahul Kumar, Prasanta Kumar Sahoo, Sandeep Kumar, Bogdan Stefan Vasile, Cristina Chircov, Jaison Darson, Mothilal Mohan, Poonam Lathiya, Jing Wang, Lizeng Gao, Kuldeep Chand Verma, Navdeep Goyal, Xiao-Lan Huang

First published in London, United Kingdom, 2022 by IntechOpen
IntechOpen is the global imprint of INTECHOPEN LIMITED, registered in England and Wales, registration number: 11086078, 5 Princes Gate Court, London, SW7 2QJ, United Kingdom

British Library Cataloguing-in-Publication Data
A catalogue record for this book is available from the British Library

Additional hard and PDF copies can be obtained from orders@intechopen.com

Iron Oxide Nanoparticles
Edited by Xiao-Lan Huang
p. cm.
Print ISBN 978-1-80355-174-6
Online ISBN 978-1-80355-175-3
eBook (PDF) ISBN 978-1-80355-176-0

Meet the editor

Dr. Xiao-Lan Huang is an independent researcher and senior environmental scientist who focuses on the biogeochemistry of soil, water, plant, and solid wastes, including phosphorus. Rachel Carson's Silent Spring inspired him to learn soil science. He holds an MSc in Soil Ecology from Nanjing Agricultural University, China, and a Ph.D. in Environmental Chemistry from the Hebrew University of Jerusalem, Israel. He also completed a postdoctoral fellowship in Oceanography at the University of Miami, USA. In 2007, Dr. Huang observed the catalytic phenomena of the phosphate ester hydrolysis in aged inorganic iron solutions and iron oxide nanomaterials under the aqueous environment with Michaelis-Menten kinetics behavior and suggested a hypothesis on the inorganic enzyme (University of Miami, 2008; RSC Advance, 2012; Astrobiology, 2018; Iron Oxide Nanoparticles: An Inorganic Phosphatase, 2019). His discovery is not only significant for organic phosphorus transformation in the current environment but also provides clues for biocatalysts and first life evolution. Dr. Huang further stated that all inorganic nanomaterials that have intrinsic enzyme-like properties can be considered inorganic enzymes due to their unique architectural features, which are comparable to the function of known enzymes (proteins) and ribozymes (RNA), in a new discovery catalog of biocatalysts. Dr. Huang is also an active participant in environmental science, as a peer reviewer for twenty international journals.

Contents

Preface

Iron oxide is one of the most abundant minerals on Earth. It presents in various nanoparticle forms that include several different phases, for example, ferrihydrite (Fh), magnetite (Fe_3O_4), maghemite (γ-Fe_2O_3), and wüstite ($Fe_{1-x}O$). Iron oxide nanoparticles are synthesized due to their innovative functions and applications. Their surface structure, shape, and size dependence make them useful for various functions. This book presents the latest developments in the research on iron oxide nanoparticles, including their synthesis, characteristics (e.g., enzyme-like activity), and applications (e.g., cancer theranostic, green energy applications). Further, the book includes additional references and perspectives in the introduction chapter regarding history, research trends, and the significance of iron architecture in nanoparticles.

Considering the intrinsic enzyme-like properties of inorganic iron oxide nanoparticles, I proposed that all inorganic nanozymes represent a new distinct class of biocatalysts (i.e., inorganic enzymes), owing to the unique architecture of the metal sites embedded in the inorganic nanomaterial. In this sense, they can be comparable to the existing categorization of enzymes (proteins) and ribozymes (RNA). Alternatively, functionalized nanomaterials that display enzyme-like properties may be referred to as "artificial biocatalysts" rather than "artificial enzymes". Metalloenzymes, in essence, are simply functionalized nanomaterials with a metal architecture as the active center, stabilized by amino acids or nucleotides through natural selection and evolution. By drawing inspiration from the understanding of protein structure, scientists may observe more enzyme-like activity in inorganic nanomaterials that have yet to be discovered. It must be emphasized that none of these inorganic nanozymes are artificial enzymes, despite claims made by many scientists. No matter what we observe or imagine, what is certain is that inorganic enzymes continue to play an important role in all the processes occurring on our planet from a biogeochemical perspective.

This book is an invaluable resource for researchers in the field of nanotechnology because it fosters a direct connection among scientists from many different disciplines, including mineralogists, biologists, chemists, engineers, geologists, agronomists, medical professionals, and environmental scientists.

Dr. Xiao-Lan Huang
Independent Researcher,
Cincinnati, OH, USA

Section 1

Introduction

Chapter 1

Introductory Chapter: Incredible Spicy Iron Oxide Nanoparticles

Xiao-Lan Huang

1. Introduction

1.1 The history of research

Iron oxide is one of the most abundant minerals on the earth, many of them can be presented as nanoparticle form, which includes several different phases, e.g., ferrihydrite (Fh), magnetite (Fe_3O_4), maghemite (γ-Fe_2O_3), and wüstite ($Fe_{1-x}O$). The structure, shape as well as size dependence make them have various functions. Moreover, all of these nanoparticles are also can be synthesized due to their innovative functions and applications. Recently the knowledge of iron oxide nanoparticles is expanding rapidly, especially after 2007. The paper numbers related to iron oxide nanoparticles were increased significantly, compared to the topic of iron oxide, and iron (**Figure 1**).

In 1989, the first article concerning Iron Oxide Nanoparticles was published in Magnetic Resonance Imaging, based on Scopus (https://www.scopus.com).

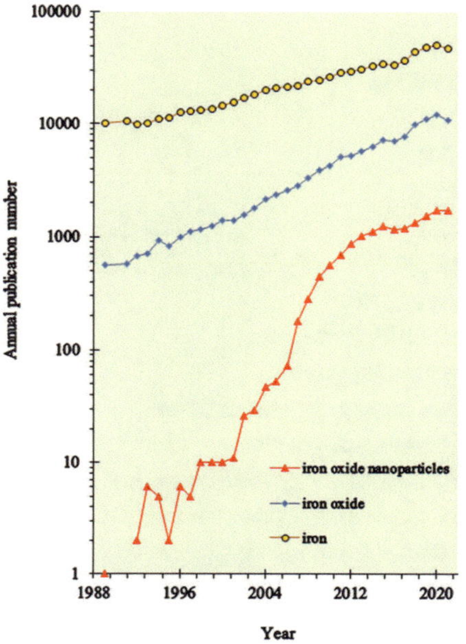

Figure 1.
The number of manuscripts related to iron, iron oxide, and iron oxide nanoparticles. Data are collected from Scopus by the end of Nov 2021.

The published article was entitled "Superparamagnetic iron oxide nanoparticles as a liver MRI contrast agent: contribution of microencapsulation to improved biodistribution" [1].

Since then, the annual number of publications related to the topic of iron oxide nanoparticles still kept a single digit from 1990 to 2001, and jumped to 3 digits in 2007, and continued to increase up to 1722 (2021). Yet, the number of manuscripts of iron, and iron oxide, has also increased. Almost 10–18% of paper on the topic "iron oxide" has been related to iron oxide nanoparticles since 2009.

It is noted that the subjects related to the topic of iron oxide nanoparticles have also increased and shifted. In the early days (1989–2001), only 68 documents in 12 years were published in the following subjects: Materials science (23%), Medicine (17%), Physics and Astronomy (15%), Chemistry (12%), and Engineering (10%), and the corresponding 5 most frequent keywords in these manuscripts are Nuclear Magnetic Resonance Imaging, Iron, Contrast Medium, Nonhuman, and Animal Experiment. More than 1700 manuscripts a year in the recent 2 years were released in up to 28 subjects (**Figure 2**), including Materials Science (19%); Chemistry (15%), Chemical Engineering (11%); Engineering (11%); Biochemistry, Genetics, and Molecular Biology (10%); Physics and Astronomy (10%); Medicine (6%); Pharmacology, Toxicology, and Pharmaceutics (5%); Environmental Science (4%); Energy (2%); Agricultural and Biological Science (1%); Immunology and Microbiology (1%); and Computer Science (1%). The other involved subjects are Mathematics, Economics, Earth and Planetary Science, Neuroscience, Health Professions, Social Science, Business, Management and Accounting, Economics, Econometrics and Finance, Dentistry, Veterinary, Psychology, Nursing, even Arts and Humanities. The corresponding 10 most frequent keywords in these manuscripts are Iron Oxides, Iron Oxide Nanoparticles, Magnetic Nanoparticles, Superparamagnetic Iron Oxide Nanoparticle, Human, Nonhuman, Animal, Chemistry, Synthesis, and Particle Size.

These manuscripts were published in 160 journals annually in 11 different languages since 2017. The top 10 journals are Nanomaterials, International Journal of

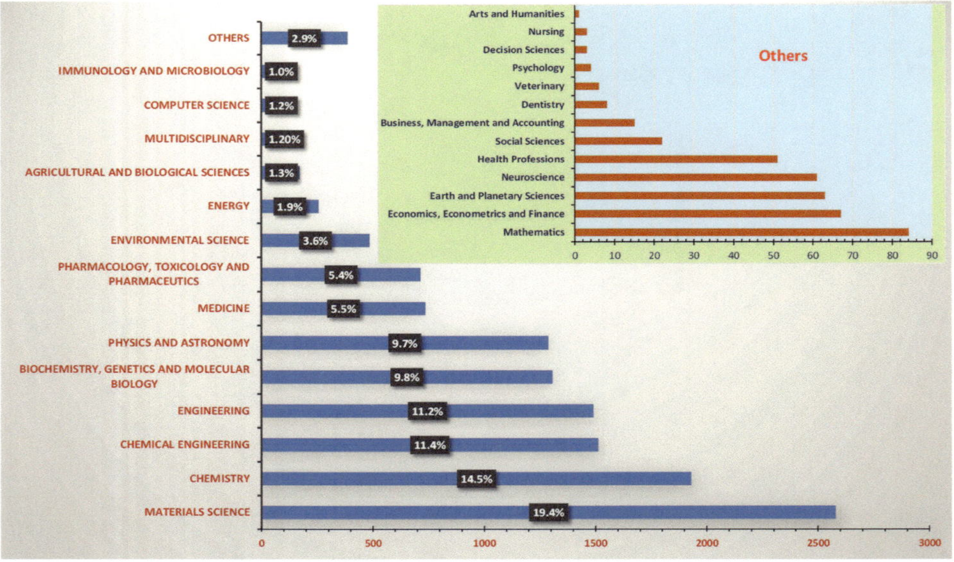

Figure 2.
The number of subjects in the manuscripts related to iron oxide nanoparticles (2017–2021). Data are collected from Scopus by the end of Nov 2021.

Nanomedicine, ACS Applied Materials and Interfaces, Scientific Reports, Journal of Magnetism and Magnetic Materials, Nanoscale, RSC Advances, International Journal of Biological Macromolecules, Materials Science and Engineering C, and Journal of Materials Chemistry B.

The support of the research related to iron oxide nanoparticles is also progressed extremely. The top 10 foundations were National Natural Science Foundation of China, National Institutes of Health, European Commission, Deutsche Forschungsgemeinschaft, Conselho Nacional de DesenvolvimentoCientífico e Tecnológico, National Research Foundation of Korea, U.S. Department of Health and Human Services, European Regional Development Fund, and Coordenação de Aperfeiçoamento de Pessoal de Nível Superior during 2017 to 2021. The top 10 countries for this area are China, the United States, India, Iran, Germany, South Korea, Spain, France, Brazil, and the United Kingdom.

2. Examples and importance of iron architecture

Many research, included several chapters in this book, specified that the importance of the crystal structure of iron oxide. The following excellent manuscripts [2–11] serve as a refresher course and basic introduction to this field, particularly the book titled "The Iron Oxides: Structure, Properties, Reactions, Occurrences, and Use" [2], the paper "Iron oxides: From molecular clusters to solid. A nice example of chemical versatility" [3], "Size-driven structural and thermodynamic complexity in iron oxide' [4], "Ordered ferrimagnetic form of ferrihydrite reveals links among structure, composition, and magnetism" [5], 'Formation, stability, and solubility of metal oxide nanoparticles: Surface entropy, enthalpy, and free energy of ferrihydrite' [6], "Crystal growth. Aqueous formation and manipulation of the iron-oxo Kegginion" [7], "Size-driven structural and thermodynamic complexity in iron oxide" [8], "Iron oxide surfaces"[9], "Unravelling the growth mechanism of the co-precipitation of iron oxide nanoparticles with the aid of synchrotron X-Ray diffraction in solution" [10] and "Ab initio thermodynamics reveals the nanocomposite structure of ferrihydrite"[11].

Here I list the top 5 citations of the original article related to iron oxide nanoparticles. They are "Intrinsic peroxidase-like activity ferromagnetic nanoparticles" [12], "Monodisperse MFe_2O_4 (M = Fe, Co, Mn) Nanoparticles" [13], "3D Nitrogen-Doped Graphene Aerogel-Supported Fe_3O_4 Nanoparticles as Efficient Electrocatalysts for the Oxygen Reduction Reaction" [14], "Synthesis of highly crystalline and monodisperse maghemite nanocrystallites without a size-selection process" [15] and "Noninvasive detection of clinically occult lymph-node metastases in prostate cancer" [16]. The first [12] and the third [14] ones were related to the new characteristics of iron oxide nanoparticles, the second [13] and the fourth [15] were related to the synthesis of iron oxide nanoparticles. The fifth [16] was related to its application. The citation number of these top 5 papers indicated that the research related to the enzyme-like activity of iron oxide nanoparticles is still hot (**Figure 3**). From these extreme samples, readers can understand why the studies and application of iron oxide nanoparticles in recent decades have become incredibly spicy.

It is interesting to note that the catalytic phenomena on phosphate ester hydrolysis, e.g., adenosine triphosphate (ATP), glucose-6-phosphate (G6P), and inorganic condensed phosphate (e.g., pyrophosphate, PP_i, and polyphosphate, poly-P) in artificial seawater was also initially observed in 2007 [17] in another independent research in the same period of Gao's work [12]. The catalysis was further inhibited by

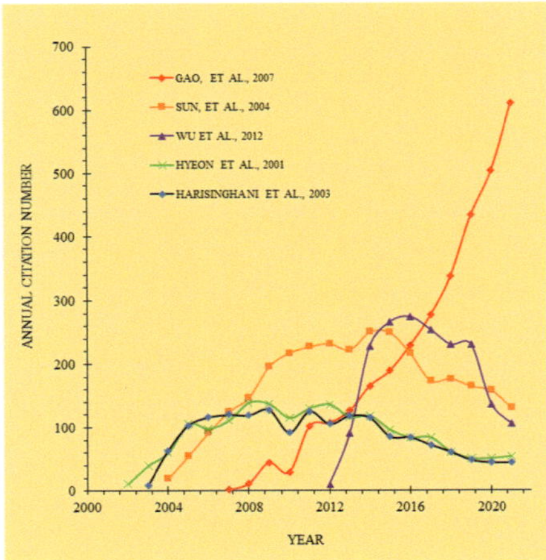

Figure 3.
The annual citation number of the top 5 papers related to iron oxide nanoparticles, data are collected from Scopus by the end of Nov 2021.

the tetrahedral oxyanions with an order of $PO_4 < MoO_4 < WO_4$, which is similar to the natural purple acid phosphatase (PAP) [18, 19]. A binuclear metal center (di-iron Fe-Fe or Fe-M (M as Mn and Zn) that produces orthophosphate due to the net transfer of the phosphoryl group to water, is essential for PAPs catalysis (**Figure 4a**) [20–24]. The author at that time claimed that the formation of diiron or polyiron with the μ-(hydr)oxo bridge through hydrolysis of iron during the aging process may contribute to the observed catalytic activity of inorganic iron oxide nanoparticles [17]. These inorganic iron oxide nanoparticles, void of protein, RNA, or any organic component might serve as an inorganic phosphoesterase. The corresponding results of phosphate ester hydrolysis, promoted by inorganic iron oxide nanoparticles with the Michaelis–Menten kinetics behavior, and hypothesis related to inorganic enzyme were published at RSC Advance [25] and Astrobiology [26] (**Figure 4**).

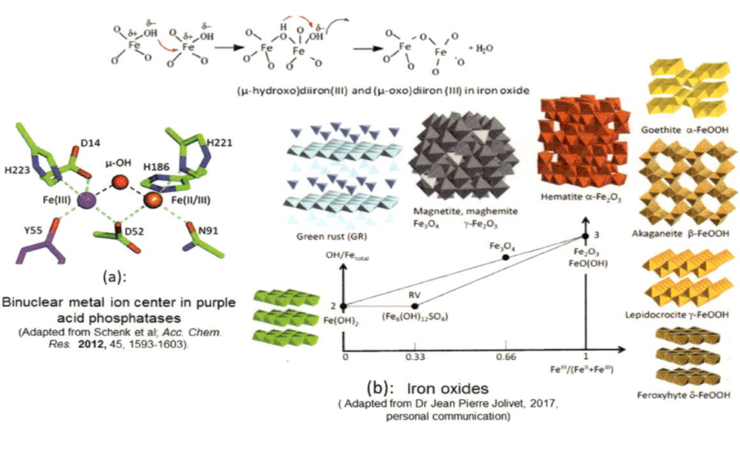

Figure 4.
Iron architecture in purple acid phosphatase and iron oxide nanoparticles, a: The μ-(hydr)oxo-bridges in purple acid phosphatase (PAP), and b: Fe-oxo-Fe structure in different iron oxides phases.

Different laboratory studies related to hydrolysis of phosphate ester further confirmed the orthophosphate releasing [27–34]. It was noted that ferrous ion (Fe II) is presumed to be the dominant form of iron in the earliest ocean [35, 36], low levels of ferric ion (Fe III) would have been produced at the ocean surface as a result of photo-oxidation, even in the absence of oxygen in the atmosphere [37, 38]. Also, precipitated ferric ion, for example in green rust in the Archaean banded iron formations, was also likely to be present in the Hadean or the early Archean [39–43]. On the other hand, inorganic condensed phosphates (e.g., PP_i), composed of orthophosphate (P_i) residue with the energy-rich phosphoanhydride bonds, are also critical to the emergence and evolution of life [44–46]. Phosphate esters, including ATP, G6P, and PP_i, is a master group of compounds involved in signaling, free-energy transduction, protein synthesis, and maintaining the integrity of the genetic material [47, 48]. Hydrolysis of these phosphate esters is related to the formation and function of the most important two biopolymers of life: RNA/DNA that encodes genetic information [49] and protein which is related to the reversible phosphorylation [50].

Similar to purple acid phosphatase, the active metal centers of most oxidoreductases in nature also comprise the transition metals, for example, horseradish peroxidase (HRP) [51] with Fe, manganese peroxidase [52], oxalate oxidase [53, 54], manganese superoxide dismutase [55] and manganese catalases [56, 57] with Mn, and haloperoxidases [58, 59] with V, all exhibit the oxo ligand structure. This unique metal architecture or metal bond in the nanoparticles might also contribute to the "intrinsic peroxidase," "intrinsic oxidase," "intrinsic superoxide dismutase," and "intrinsic catalase" feature from different inorganic metal oxides nanoparticles [12, 60–69]. It was noted that some PAPs also have activity of peroxidases [70]. Meanwhile, scientists have demonstrated that inorganic V_2O_5 nanomaterials have haloperoxidase-like activity in the presence of substrates such as Br- and H_2O_2 [68, 71]. The activity is due to the nanostructure of vanadium pentoxide, not the free vanadate in solution from leaching processes [68]. The nanowires of V_2O_5 are stacked in the [001] direction while their wires extend in the [100] direction, and a view of the (110) plane reveals extraordinary similarities with the haloperoxidase-active sites. Consequently, small amounts of hypobromous acid are continuously produced in the ocean to kill or inhibit the bacteria around the V_2O_5/paint [71]. The suggested mechanisms of the inorganic haloperoxidase, are also similar to those of natural vanadium haloperoxidase [58, 59]. Such V_2O_5/paint has been successfully applied to combat marine biofouling, i.e., the colonization of small marine microorganisms on the surfaces of ships that are directly exposed to seawater [71]. Like the nanoparticles of iron oxide and vanadium pentoxide, the intrinsic sulfite oxidase activity of molybdenum trioxide nanoparticles is also due to the oxo ligand of Mo [72], as revealed in the metal center of sulfite oxidase [73, 74]. The work using the in situ Raman spectroscopy on the changes of V-oxo (V=O) bond in the different V_2O_5 nanomaterials during H_2O_2 catalysis cycle further indicated that the catalytic characteristics in these metal oxide nanoparticles are due to the distinctive crystal structure or metal bond, i.e., not merely the simple surface area or particle size of the nanoparticles [75]. The different shapes and sizes of Fe_3O_4 nanocrystals and $MnFe_2O_4$ nanoparticles also demonstrated that the key role of metal oxides structure for their catalytic activity [76, 77], supported by the density functional theory calculation as well [78].

Inorganic nanomaterials with the enzyme-like activity were not generated by artificial synthesis processing, but their metal architecture in nanoparticles. Such catalytic kinetics also can be described by Michaelis–Menten equations, the same as the natural enzyme (protein and ribozymes) [79–82] in modern biochemistry. Inorganic iron oxide nanoparticles from ferritin and magnetosomes are also already confirmed to have such enzyme-like activity [83–87]. This further validated that these inorganic iron oxide nanoparticles with enzyme-like activity are inorganic enzymes [25, 26, 88], not artificial enzymes [89, 90]. Transition metal sulfites, even selenide nanoparticles were anticipated to form due to a reducing atmosphere and ocean in the early earth environment [35, 91]. Greigite (Fe_7S_8), as an example, was detected in the laboratory under the simulated early Earth hydrothermal conditions [92]. Some of them were documented to have "intrinsic peroxidase" "intrinsic superoxide dismutase," and/or "intrinsic catalase" activity, e.g., Fe_3S_4 [93], Fe_7S_8 [94], CuS [95], CuZnFeS [96], CdS [97], MoS_2 [98], WS_2 [99], FeS [100], FeSe [100], α-MnSe [101], $MoSe_2$ [102], and $CoSe_2$ [103]. It is not strange since many iron–sulfur proteins contain a cluster of multinuclear iron and inorganic sulfide, where the irons are coordinated by protein amino acid residues and sulfides with such function [104–106], e.g., alkyl hydroperoxide reductase with a disulfide bond structure [107]. Glutathione peroxidase with the selenocysteine as catalytic active center [108] was reported to scavenge hydrogen peroxide. These findings provide solid evidence that these biocatalysts can be presented before the protein and RNA world, and thereby offer a solution to the "chicken and egg" at life's beginnings [109, 110]. All of these inorganic enzyme activities, due to their specialized crystal structure including but not limited to their surface area, do support the metabolism-first hypothesis, not the replicator-first scenario [111, 112], and are also to be considered highly important in the context of new theories about the emergence of life [113–115].

Actually, iron oxide nanoparticles are very common nanoparticles present in the soil, sediment, water, even air dust in our current earth environment [116–120]. It is very reasonable to expect that these iron oxide nanoparticles, as inorganic phosphatase, play a significant role for organic phosphorus in the water and soil nowadays too. This is an underestimated pathway for organic phosphorus transformation from the view of biogeochemistry [31, 121–126], since different phosphate esters, especially monoesters, are the main components in the dissolved organic phosphorus in soils [127–130] and waters [131–134]. Furthermore, several recent investigations already confirmed that iron-rich nanoparticles (< 20 nm) are the main carriers of phosphorus in forest streams and soil solutions [135–139]. Keep in mind that many environmental factors also impact the activity of the catalysis of these inorganic enzymes, including the inorganic phosphatases [25, 26, 88]. The activity of inorganic phosphatases can be inhibited by some small organic molecules, e.g., citrate acid, due to the iron complex formation [26, 88]. The interaction of organic matter and iron oxide nanoparticles is very complex [140–142], the catalytic capacity, on the other hand, can also be enhanced when metal oxide nanoparticles surface modified with some especially organic ligand, e.g., glutathione, dendrimer, DNA, and protein, based on the progress of nanozyme and metal–organic frameworks from the view of bioengineering [143–147].

The fact of orthophosphate release from the hydrolysis of organic phosphorus, promoted by the iron oxide nanoparticles, transfigures the basic assumptions on the iron-sorbed phosphorus for the Redfield stoichiometry [148], which is a fundamental feature in the understanding of the biogeochemical cycles of the oceans [149–153]. Such new discovery related to the role of iron oxide nanoparticles significantly impacts the carbon and nutrient fluxes in global circulation models [154–156]. It also further challenges the notion of ocean iron fertilization as a potential method of removing atmosphere CO_2 technologies to reduce the temperature increase for the coming climate changes [157–163]. The implications of its catalytic activity related to organic phosphorus transformation are not limited to the emergence of life, phytoavailability of phosphorus, but also go to global carbon and nutrient fluxes, and climate changes. The incredible spicy iron oxide nanoparticles!

3. Additional references and perspectives

Iron oxide nanoparticles are also found in our everyday life, even if we don't realize it. They are bioactive materials and may perform various biological functions in life activity, especially related to reactive oxygen species. They can be used as a valuable tool in cancer therapeutic application, which is presented in this book. The incredible spicy iron oxide nanoparticles!

I will argue that the applications of iron oxide nanoparticles do not stop at biomedicine. In addition to being extremely useful for green energy applications such as supercapacitors and hydroelectric cells, these inorganic iron oxide nanoparticles are also used extensively in the radio frequency and microwave applications discussed in this book. The incredible spicy iron oxide nanoparticles!

I do appreciate the contributions of all authors of this book. Indeed, this book has covered much significant progress of iron oxide nanoparticles, especially its medical and green energy applications. However, some very important components related to iron oxide nanoparticles in nature are still missing. I encourage readers to look up the following reviews to have a BIG picture of iron oxide nanoparticles.

Guo and Barnard [117] focused their review, titled "Naturally occurring iron oxide nanoparticles: morphology, surface chemistry and environmental stability", on the several phases (ferrihydrite, goethite, hematite, magnetite, maghemite, lepidocrocite, akaganéite, and schwertmannite) commonly found in water, soils, and sediments. Their functions in various aspects are closely related to their shapes, sizes, and thermodynamic surroundings. Phase transformations and the relative abundance are sensitive to changes in environmental conditions.

Braunschweig *et al.* [164], in their study "Iron oxide nanoparticles in geomicrobiology: From biogeochemistry to bioremediation", examine a number of factors influencing the microbial reactivity of Fe oxides, including particle size, solubility, ferrous iron, crystal structure, and organic molecules. It highlights the differences between natural and synthetic Fe oxides.

By using the title "Iron solubility, colloids, and their impact on iron oxide formation from solution", Baumgartner and Faivre [165] started their review from the iron solubility, and its speciation in water as well as phase transformation processes from solution to solid and vice versa. It included experimental findings and theoretical concepts of iron (oxyhydr)oxide dissolution and formation in aqueous solution: hydrolysis, nanoparticle size-dependent aspects of solubility and particle formation pathways, nucleation, growth, and oriented attachment.

A article by Claudio *et al.* [166] focused on the importance of iron oxide nanoparticles in the soil, especially related to soil fertility, plant nutrition, and the interaction of phosphorus, sulfate, molybdate, and pollutants (arsenic or chromium) from the traditional view of soil chemistry with the title " Iron Oxide Nanoparticles in Soils: Environmental and Agronomic Importance". While Vindedahl *et al.* [141], under the title " Organic matter and iron oxide nanoparticles: Aggregation, interactions, and reactivity", addressed the chemistry of iron oxide nanoparticles in aqueous environments, e.g., the effect of pH, organic matter sorption, and solid-state transformations. Both can help readers understand the fate, transport, and chemical behavior of nanoparticles in complex environments.

Nanozymes are defined by Wei and Wang as nanomaterials with enzyme-like properties [89, 167]. In comparison to a previous literature review on this topic [89] (2013, cited 302 documents), "Nanomaterials with enzyme-like characteristics (nanozymes): Next-generation artificial enzymes (II)" [167] cites 1212 documents. It provides a general overview of this research field, including the type of nanozymes and their representative nanomaterials, catalytic mechanisms as well as broad applications. The challenges and the directions for advancing nanozyme research were also suggested by the authors. However, the concept of "artificial enzymes" or "enzyme mimics" cannot be applied to these inorganic nanozymes. Specifically, inorganic nanomaterials possessing enzyme-like activity are attributed to their unique architecture, as previously noted. Such inorganic nanomaterials architecture is not created artificially but rather is intrinsic to the mineral crystal itself. Scientists only observed enzyme-like activities in nanomaterials but did not create these feathers. To distinguish inorganic nanozymes from the previously existing classification of enzymes (proteins) and ribozymes (RNAs) as biocatalysts, I proposed classifying them as inorganic enzymes.

Malhotra *et al.*, with the title "Potential Toxicities of Iron Oxide Magnetic Nanoparticles: A Review" [168], addressed the safety of engineering iron oxide nanoparticles for different applications. Many factors, including their surface to volume ratio, chemical composition, size, and dosage, retention in the body, immunogenicity, organ-specific toxicity, breakdown, and elimination from the body can impact their toxicity. More research is needed to further assess its safety.

Our book, together with these additional manuscripts and reviews, aims to provide an active source in the field of nanoscience while creating a bridge between scientists and engineers working in the fields of mineralogy, biology, chemistry, geology, agronomy, medicine, environmental sciences, as well as the green energy industry.

Author details

Xiao-Lan Huang
Independent Researcher, United States of America

*Address all correspondence to: xiaolan.huang@ymail.com

References

[1] Pouliquen D, Perdrisot R, Ermias A, Akoka S, Jallet P, Le Jeune JJ. Superparamagnetic iron oxide nanoparticles as a liver MRI contrast agent: Contribution of microencapsulation to improved biodistribution. Magnetic Resonance Imaging. 1989;7(6):619-627. DOI: 10.1016/0730-725x(89)90530-4

[2] Cornell RM, Schwertmann U. The Iron Oxides: Structure, Properties, Reactions, Occurrences and Uses. 2nd ed. Weinheim, Germany: Wiley-VCH; 2003. DOI: 10.1002/3527602097

[3] Jolivet JP, Tronc E, Chanéac C. Iron oxides: From molecular clusters to solid. A nice example of chemical versatility. ComptesRendus – Geoscience. 2006;**338**:488-497. DOI: 10.1016/j.crte.2006.04.014

[4] Navrotsky A, Mazeina L, Majzlan J. Size-driven structural and thermodynamic complexity in iron oxides. Science. 2008;**319**(5870):1635-1638. DOI: 10.1126/science.1148614

[5] Michel FM, Barrón V, Torrent J, Morales MP, Serna CJ, Boily JF, et al. Ordered ferrimagnetic form of ferrihydrite reveals links among structure, composition, and magnetism. Proceedings of the National Academy of Sciences of the United States of America. 2010;**107**(7):2787-2792. DOI: 10.1073/pnas.0910170107

[6] Hiemstra T. Formation, stability, and solubility of metal oxide nanoparticles: Surface entropy, enthalpy, and free energy of ferrihydrite. Geochimica et Cosmochimica Acta. 2015;**158**:179-198. DOI: 10.1016/j.gca.2015.02.032

[7] Sadeghi O, Zakharov LN, Nyman M. Crystal growth. Aqueous formation and manipulation of the iron-oxo Keggin ion. Science. 2015;**347**(6228):1359-1362. DOI: 10.1126/science.aaa4620

[8] Weatherill JS, Morris K, Bots P, Stawski TM, Janssen A, Abrahamsen L, et al. Ferrihydrite formation: The role of Fe13 keggin clusters. Environment Science & Technology. 2016;**50**(17):9333-9342. DOI: 10.1021/acs.est.6b02481

[9] Parkinson GS. Iron oxide surfaces. Surface Science Reports. 2016;**71**(1):272-365. DOI: 10.1016/j.surfrep.2016.02.001

[10] LaGrow AP, Besenhard MO, Hodzic A, Sergides A, Bogart LK, Gavriilidis A, et al. Unraveling the growth mechanism of the co-precipitation of iron oxide nanoparticles with the aid of synchrotron X-ray diffraction in solution. Nanoscale. 2019;**11**(14):6620-6628. DOI: 10.1039/c9nr00531e

[11] Sassi M, Chaka AM, Rosso KM. Ab initio thermodynamics reveals the nanocomposite structure of ferrihydrite. Communications Chemistry. 2021;**4**:134. DOI: 10.1038/s42004-021-00562-7

[12] Gao L, Zhuang J, Nie L, Zhang J, Zhang Y, Gu N, et al. Intrinsic peroxidase-like activity of ferromagnetic nanoparticles. Nature Nanotechnology. 2007;**2**(9):577-583. DOI: 10.1038/nnano.2007.260

[13] Sun S, Zeng H, Robinson DB, Raoux S, Rice PM, Wang SX, et al. Monodisperse MFe_2O_4 (M = Fe, Co, Mn) nanoparticles. Journal of the American Chemical Society. 2004;**126**(1):273-279. DOI: 10.1021/ja0380852

[14] Wu ZS, Yang S, Sun Y, Parvez K, Feng X, Müllen K. 3D nitrogen-doped graphene aerogel-supported Fe_3O_4

nanoparticles as efficient electrocatalysts for the oxygen reduction reaction. Journal of the American Chemical Society. 2012;**134**(22):9082-9085. DOI: 10.1021/ja3030565

[15] Hyeon T, Lee SS, Park J, Chung Y, Na HB. Synthesis of highly crystalline and monodisperse maghemite nanocrystallites without a size-selection process. Journal of the American Chemical Society. 2001;**123**(51):12798-12801. DOI: 10.1021/ja016812s

[16] Harisinghani MG, Barentsz J, Hahn PF, Deserno WM, Tabatabaei S, van de Kaa CH, et al. Noninvasive detection of clinically occult lymph-node metastases in prostate cancer. The New England Journal of Medicine. 2003;**348**(25):2491-2499. DOI: 10.1056/NEJMoa022749

[17] Xiao-Lan H, Zhang J-Z. Sediment-water exchange of dissolved organic phosphorus in Florida bay. In: CIMAS Annual Report 2007-2008, Cooperative Institute for Marine and Atmospheric Studies. Miami, FL: University of Miami; 2008. pp. 72-73. Available from: https://cimas.rsmas.miami.edu/_assets/pdf/annual-reports/2008_annual_report.pdf

[18] Lim JS, Aquino MAS, Sykes AG. Mechanistic studies on the reactions of molybdenum(VI), tungsten(VI), vanadium(V), and arsenic(V) tetraoxo anions with the FeIIFeIII form of purple acid phosphatase from porcine uteri (uteroferrin). Inorganic Chemistry. 1996;**35**:614-618. DOI: 10.1021/ic950139o

[19] Elliott TW, MitiÄ N, Gahan LR, Guddat LW, Schenk G. Inhibition studies of purple acid phosphatases: Implications for the catalytic mechanism. Journal of the Brazilian Chemical Society. 2006;**17**:1558-1565. DOI: 10.1590/S0103-50532006000800011

[20] Schenk G, Carrington LE, Hamilton SE, Jersey JD, Guddat LW. Crystallization and preliminary X-ray diffraction data for a purple acid phosphatase from sweet potato. Acta Crystallographica Section D: Structural Biology. 1999;**55**:2051-2052. DOI: 10.1107/S0907444999012597

[21] Schenk G, Mitić N, Hanson GR, Comba P. Purple acid phosphatase: A journey into the function and mechanism of a colorful enzyme. Coordination Chemistry Reviews. 2013;**257**:473-482. DOI: 10.1016/j.ccr.2012.03.020

[22] Sträter N, Lipscomb WN. Two-metal ion catalysis in enzymatic acyl- and phosphoryl-transfer reactions. Angewandte Chemie International Edition. 1996;**35**:2024-2055. DOI: 10.1002/anie.199620241

[23] Lindqvist Y, Johansson E, Kaija H, Vihko P, Schneider G. Three-dimensional structure of a mammalian purple acid phosphatase at 2.2 Å resolution with a μ-(hydr)oxo bridged di-iron center. Journal of Molecular Biology. 1999;**291**:135-147. DOI: 10.1006/jmbi.1999.2962

[24] Mitić N, Smith SJ, Neves A, Guddat LW, Gahan LR, Schenk G. The catalytic mechanisms of binuclear metallohydrolases. Chemical Reviews. 2006;**106**:3338-3363. DOI: 10.1021/cr050318f

[25] Xiao-Lan H, Zhang J-Z. Hydrolysis of glucose-6-phosphate in aged, acid-forced hydrolysed nanomolar inorganic iron solution — An inorganic biocatalyst? RSC Advance. 2012;**2**:199-208. DOI: 10.1039/c1ra00353d

[26] Huang X-L. Hydrolysis of phosphate esters catalyzed by inorganic iron oxide nanoparticles acting as biocatalysts. Astrobiology. 2018;**18**(3):294-310. DOI: 10.1089/ast.2016.1628

[27] Olsson R, Giesler R, Loring JS, Persson P. Adsorption, desorption, and surface-promoted hydrolysis of glucose-1-phosphate in aqueous goethite (alpha-feooh) suspensions. Langmuir. 2010;**26**:18760-18770. DOI: 10.1021/la1026152

[28] Mäkie P, Westin G, Persson P, Österlund L. Adsorption of trimethyl phosphate on maghemite, hematite, and goethite nanoparticles. The Journal of Physical Chemistry A. 2011;**115**(32):8948-8959. DOI: 10.1021/jp201065w

[29] Persson P, Andersson T, Nelson H, Sjöberg S, Giesler R, Lövgren L. Surface complexes of monomethyl phosphate stabilized by hydrogen bonding on goethite (α-FeOOH) nanoparticles. Journal of Colloid and Interface Science. 2012;**386**(1):350-358. DOI: 10.1016/j.jcis.2012.07.042

[30] Lü C, Yan D, He J, Zhou B, Li L, Zheng Q. Environmental geochemistry significance of organic phosphorus: An insight from its adsorption on iron oxides. Applied Geochemistry. 2017;**84**:52-60. DOI: 10.1016/j.apgeochem.2017.05.026

[31] Wan B, Huang R, Diaz JM, Tang Y. Rethinking the biotic and abiotic remineralization of complex phosphate molecules in soils and sediments. Science of the Total Environment. 2022:**833**:155187. DOI:10.1016/j.scitotenv.2022.155187

[32] Fang Y, Kim E, Strathmann TJ. Mineral- and base-catalyzed hydrolysis of organophosphate flame retardants – A potential major fate-controlling sink in soil and aquatic environments. Environmental Science & Technology. 2018;**52**:1997-2006. DOI: 10.1021/acs.est.7b0591

[33] Klein AR, Bone SE, Bakker E, Chang Z, Aristilde L. Abiotic phosphorus recycling from adsorbed ribonucleotides on a ferrihydrite-type mineral: Probing solution and surface species. Journal of Colloid and Interface Science. 2019;**547**:171-182. DOI: 10.1016/j.jcis.2019.03.086

[34] Wan B, Yang P, Jung H, Zhu M, Diaz JM, Tang Y. Iron oxides catalyze the hydrolysis of polyphosphate and precipitation of calcium phosphate minerals. Geochemica et Cosmochimica Acta. 2021;**305**:49-65. DOI: 10.1016/j.gca.2021.04.031

[35] Holland HD. The geologic history of sea water—An attempt to solve the problem. Geochimica et Cosmochimica Acta. 1972;**36**:637-651. DOI: 10.1016/0016-7037(72)90108-1

[36] Raiswell R, Canfield DE. Section 2. The iron cycle: Biogeochemistry and mineralogy. Geochemical Perspectives. 2012;**1**(1):11-18. DOI: 10.7185/geochempersp.1.1

[37] Braterman PS, Cairns-Smith AG, Sloper RW. Photooxidation of hydrated Fe^{2+} significance for banded iron formations. Nature. 1983;**303**(5913):1 63-164. DOI: 10.1038/303163a0

[38] Bekker A, Planavsky N, Rasmussen B, Krapež B, et al. Iron formations: Their origins and implications for ancient seawater chemistry. In: Holland H, Turekian K, editors. Treatise on Geochemistry. 2nd ed. Netherlands: Elsevier Inc.; 2013. pp. 561-628. DOI: 10.1016/B978-0-08-095975-7.00719-1

[39] Génin J-MR, Mills SJ, Christy AG, Guérin O, Herbillon AJ, Kuzmann E, et al. Mössbauerite, $Fe^{3+}{}_6O_4(OH)_8[CO_3]\cdot 3H_2O$, the fully oxidized 'green rust' mineral from Mont saint-Michel Bay, France. Mineralogical Magazine. 2014;**78**(2):447-465. DOI: 10.1180/minmag.2014.078.2.14

[40] Nisbet EG, Sleep NH. The habitat and nature of early life. Nature. 2001; **409**(6823):1083-1091. DOI: 10.1038/35059210

[41] Posth NR, Canfield DE, Kappler A. Biogenic Fe(III) minerals: From formation to diagenesis and preservation in the rock record. Earth-Science Reviews. 2014;**135**:103-121. DOI: 10.1016/j.earscirev.2014.03.012

[42] Shibuya T, Russell M, Takai K. Free energy distribution and chimney minerals in Hadean submarine alkaline vent systems; importance of iron redox reactions under anoxic condition. Geochimica et Cosmochimica Acta. 2016;**175**:1-19. DOI: 10.1016/j.gca.2015.11.021

[43] Konhauser KO, Planavsky NJ, Hardisty DS, et al. Iron formations: A global record of Neoarchaean to Palaeoproterozoic environmental history. Earth-Science Reviews. 2017;**172**:140-177. DOI: 10.1016/j.earscirev.2017.06.012

[44] Baltscheffsky H. Inorganic pyrophosphate and the origin and evolution of biological energy transformation. In: Chemical Evolution and the Origin of Life, Proceedings of the International Conference on the Origin of Life. Amsterdam: North-Holland Pub; 1971. pp. 466-474

[45] Kornberg A, Rao NN, Ault-Riché D. Inorganic polyphosphate: A molecule of many functions. Annual Review of Biochemistry. 1999;**68**:89-125. DOI: 10.1146/annurev.biochem.68.1.89

[46] Holm NG, Baltscheffsky H. Links between hydrothermal environments, pyrophosphate, Na^+, and early evolution. Origins of Life and Evolution of Biospheres. 2011;**41**:483-493. DOI: 10.1007/s11084-011-9235-4

[47] Bowler MW, Cliff MJ, Waltho JP, Blackburn GM. Why did nature select phosphate for its dominant roles in biology? New Journal of Chemistry. 2010;**34**:784-794. DOI: 10.1039/B9NJ00718K

[48] Kamerlin S, Sharma P, Prasad R, Warshel A. Why nature really chose phosphate. Quarterly Reviews of Biophysics. 2013;**46**(1):1-132. DOI: 10.1017/S0033583512000157

[49] Westheimer FH. Why nature chose phosphates. Science. 1987;**235**:1173-1178. DOI: 10.1126/science.2434996

[50] Hunter T. Why nature chose phosphate to modify proteins. Philosophical Transactions of the Royal Society B: Biological Sciences. 2012;**367**:2513-2516. DOI: 10.1098/rstb.2012.0013

[51] Gajhede M, Schuller DJ, Henriksen A, Smith AT, Poulos TL. Crystal structure of horseradish peroxidase C at 2.15 a resolution. Nature Structural & Molecular Biology. 1997;**4**(12):1032-1038. DOI: 10.1038/nsb1297

[52] Sundaramoorthy M, Kishi K, Gold MH, Poulos TL. The crystal structure of manganese peroxidase from Phanerochuetechrysosporium at 2.06-a resolution. Journal of Biological Chemistry. 1994;**269**:32759-32767. DOI: 10.1016/S0021-9258(20)30056-9

[53] Requena L, Bornemann S. Barley (Hordeum vulgare) oxalate oxidase is a manganese-containing enzyme. Biochemical Journal. 1999;**343**(1):185-190. DOI: 10.1042/bj3430185

[54] Opaleye O, Rose R-S, Whittaker MM, Woo E-J, Whittaker JW, Pickersgill RW. Structural and spectroscopic studies shed light on the mechanism of oxalate oxidase. Journal of Biological Chemistry.

2006;**281**(10):6428-6433. DOI: 10.1074/jbc.M510256200

[55] Stallings WC, Pattridge KA, Strong RK, Ludwig ML. The structure of manganese superoxide dismutase from Thermus thermophilus HB8 at 2.4-a resolution. Journal of Biological Chemistry. 1985;**260**(30):16424-16432. DOI: 10.1016/S0021-9258(17)36254-3

[56] Barynin VV, Whittaker MM, Antonyuk SV, Lamzin VS, Harrison PM, Artymiuk PJ, et al. Crystal structure of manganese catalase from lactobacillus plantarum. Structure. 2001;**9**(8):725-738. DOI: 10.1016/s0969-2126(01)00628-1

[57] Dismukes GC. Manganese enzymes with binuclear active sites. Chemical Reviews. 1996;**96**:2909-2926. DOI: 10.1021/cr950053c

[58] Messerschmidt A, Wever R. X-ray structure of a vanadium-containing enzyme: Chloroperoxidase from the fungus Curvulariainaequalis. Proceedings of the National Academy of Sciences of the United States of America. 1996;**93**:392-396. DOI: 0.1073/pnas.93.1.392

[59] Chen ZL. Recent development of biomimetic halogenation inspired by vanadium dependent haloperoxidase. Coordination Chemistry Reviews. 2022;**457**:214404. DOI: 10.1016/j.ccr.2021.214404

[60] Chaudhari KN, Chaudhari NK, Yu J-S. Peroxidase mimic activity of hematite iron oxides (α-Fe_2O_3) with different nanostructures. Catalysis Science & Technology. 2012;**2**:119-124. DOI: 10.1039/C1CY00124H

[61] Chen Z, Yin JJ, Zhou YT, Zhang Y, Song L, Song M, et al. Dual enzyme-like activities of iron oxide nanoparticles and their implication for diminishing cytotoxicity. ACS Nano. 2012;**6**(5):4001-4012. DOI: 10.1021/nn300291r

[62] Liu X, Wang Q, Zhao H, Zhang L, Su Y, Lv Y. BSA-templated MnO_2 nanoparticles as both peroxidase and oxidase mimics. Analyst. 2012;**137**(19):4552-4558. DOI: 10.1039/c2an35700c

[63] Zhang X, Huang Y. Evaluation of the antioxidant activity of phenols and tannic acid determination with Mn_3O_4 nano-octahedrons as an oxidase mimic. Analytical Methods. 2015;**7**(20):8640-8646. DOI: 10.1039/c5ay01732g

[64] Vernekar AA, Das T, Ghosh S, Mugesh G. A remarkably efficient $MnFe_2O_4$-based oxidase nanozyme. Chemistry: An Asian Journal. 2016;**11**:72-76. DOI: 10.1002/asia.201500942

[65] Su L, Qin W, Zhang H, Rahman ZU, Ren C, Ma S, et al. The peroxidase/catalase-like activities of MFe_2O_4 (M=Mg, Ni, Cu) MNPs and their application in colorimetric biosensing of glucose. Biosensors and Bioelectronics. 2015;**63**:384-391. DOI: 10.1016/j.bios.2014.07.048

[66] Singh N, Savanur MA, Srivastava S, D'Silva P, Mugesh G. A redox modulatory Mn_3O_4 nanozyme with multi-enzyme activity provides efficient cytoprotection to human cells in a Parkinson's disease model. Angewandte Chemie International Edition in English. 2017;**56**(45):14267-14271. DOI: 10.1002/anie.201708573

[67] Chong Y, Liu Q, Ge C. Advances in oxidase-mimicking nanozymes: Classification, activity regulation and biomedical applications. Nano Today. 2021;**37**:101076. DOI: 10.1016/j.nantod.2021.101076

[68] André R, Natálio F, Humanes M, Leppin J, Heinze K, Wever R, et al. V_2O_5

nanowires with an intrinsic peroxidase-like activity. Advanced Functional Materials. 2011;**21**(3):501-509. DOI: 10.1002/adfm.201001302

[69] Vernekar A, Sinha D, Srivastava S, et al. An antioxidant nanozyme that uncovers the cytoprotective potential of vanadia nanowires. Nature Communications. 2014;**5**:5301. DOI: 10.1038/ncomms6301

[70] Hayman AR, Cox TM. Purple acid phosphatase of the human macrophage and osteoclast. Characterization, molecular properties, and crystallization of the recombinant di-iron-oxo protein secreted by baculovirus-infected insect cells. Journal of Biological Chemistry. 1994;**269**(2):1294-1300. DOI: 10.1016/S0021-9258(17)42257-5

[71] Natalio F, André R, Hartog AF, Stoll B, Jochum KP, Wever R, et al. Vanadium pentoxide nanoparticles mimic vanadium haloperoxidases and thwart biofilm formation. Nature Nanotechnology. 2012;**7**:530-535. DOI: 10.1038/nnano.2012.91

[72] Ragg R, Natalio F, Tahir MN, Janssen H, Kashyap A, Strand D, et al. Molybdenum trioxide nanoparticles with intrinsic sulfite oxidase activity. ACS Nano. 2014;**8**(5):5182-5189. DOI: 10.1021/nn501235j

[73] Feng C, Tollin G, Enemark JH. Sulfite oxidizing enzymes. Biochimica et Biophysica Acta. 2007;**1774**(5):527-539. DOI: 10.1016/j.bbapap.2007.03.006

[74] Schwarz G, Mendel RR, Ribbe MW. Molybdenum cofactors, enzymes and pathways. Nature. 2009;**460**(7257): 839-847. DOI: 10.1038/nature08302

[75] Ghosh S, Roy P, Karmodak N, Jemmis ED, Mugesh G. Nanoisozymes: Crystal-facet-dependent enzyme-mimetic activity of V_2O_5 nanomaterials. Angewandte Chemie International Edition in English. 2018;**57**(17):4510-4515. DOI: 10.1002/anie.201800681

[76] Liu S, Lu F, Xing R, Zhu J-J. Structural effects of Fe_3O_4 nanocrystals on peroxidase-like activity. Chemistry - A European Journal. 2011;**17**(2):620-625. DOI: 10.1002/chem.201001789

[77] Peng Y, Wang Z, Liu W, et al. Size- and shape-dependent peroxidase-like catalytic activity of $MnFe_2O_4$ nanoparticles and their applications in highly efficient colorimetric detection of target cancer cells. Dalton Transactions. 2015;**44**(28):12871-12877. DOI: 10.1039/c5dt01585e

[78] Shen XM, Wang ZZ, Gao XF, Zhao YL. Density functional theory-based method to predict the activities of nanomaterials as peroxidase mimics. ACS Catalysis. 2020;**10**(21):12657-12665. DOI: 10.1021/acscatal.0c03426

[79] Jencks WP. Catalysis in Chemistry and Enzymulogy. New York: McGraw Hill; 1969

[80] Kohel JRE. The enzyme theory and the origin of biochemistry. Isis. 1973;**64**(2):181-196. DOI: 10.1086/351080

[81] Zaug AJ, Cech TR. The intervening sequence RNA of Tetrahymena is an enzyme. Science. 1986;**231**(4737):470-475. DOI: 10.1126/science.3941911

[82] Narlikar GJ, Herschlag D. Mechanistic aspects of enzymatic catalysis: Lessons from comparison of RNA and protein enzymes. Annual Review of Biochemistry. 1997;**66**:19-59. DOI: 10.1146/annurev.biochem.66.1.19

[83] Arapova GS, Eryomin AN, Metelitza DI. Catalytic peroxidase-like activity of an iron-containing crystallite

isolated from ferritin in aqueous solution and reversed micelles of aerosol OT in heptane. Russian Journal of Bioorganic Chemistry. 1999;**25**(5):369-376

[84] Hu L, Song T, Ma Q, Chen C, Pan W, Xie C, Nie L, Yang W. Bacterial magnetic nanoparticles as peroxidase mimetics and application in immunoassay. AIP Conference Proceedings. 2010;**1311**(1):369-374. DOI: 10.1063/1.3530039

[85] Pan Y, Li N, Mu J, Zhou R, Xu Y, Cui D, et al. Biogenic magnetic nanoparticles from Burkholderia sp. YN01 exhibiting intrinsic peroxidase-like activity and their applications. Applied Microbiology and Biotechnology. 2014;**99**:703-715. DOI: 10.1007/s00253-014-5938-6

[86] Jiang B, Fang L, Wu K, Yan X, Fan K. Ferritins as natural and artificial nanozymes for theranostics. Theranostics. 2020;**10**(2):687-706. DOI: 10.7150/thno.39827

[87] Zhang J, Li S, Xie N, Nie G, Tang A, Zhang XE, et al. A natural nanozyme in life is found: The iron core within ferritin shows superoxide dismutase catalytic activity. Science China. Life Sciences. 2021;**64**(8):1375-1378. DOI: 10.1007/s11427-020-1865-2

[88] Huang X-L. Iron oxide nanoparticles: An inorganic phosphatase. In: Sinha I, Shukla M, editors. Nanocatalysts. London: IntechOpen; 2019. DOI: 10.5772/intechopen.82650

[89] Wei H, Wang E. Nanomaterials with enzyme-like characteristics (nanozymes): Next-generation artificial enzymes. Chemistry Society Review. 2013;**42**(14):6060-6093. DOI: 10.1039/c3cs35486e

[90] Lin Y, Ren J, Qu X. Catalytically active nanomaterials: A promising candidate for artificial enzymes. Accounts of Chemical Research. 2014;**47**(4):1097-1105. DOI: 10.1021/ar400250z

[91] Trail D, Watson EB, Tailby ND. The oxidation state of Hadean magmas and implications for early Earth's atmosphere. Nature. 2011;**480**(7375):79-82. DOI: 10.1038/nature10655

[92] White LM, Bhartia R, Stucky GD, Kanik I, Russell MJ. Mackinawite and greigite in ancient alkaline hydrothermal chimneys: Identifying potential key catalysts for emergent life. Earth and Planetary Science Letters. 2015;**430**: 105-114. DOI: 10.1016/j.epsl.2015.08.013

[93] Ding C, Yan Y, Xiang D, et al. Magnetic Fe_3S_4 nanoparticles with peroxidase-like activity, and their use in a photometric enzymatic glucose assay. Microchimica Acta. 2016;**183**:625-631. DOI: 10.1007/s00604-015-1690-6

[94] Yao WT, Zhu HZ, Li WG, Yao HB, Wu YC, Yu SH. Intrinsic peroxidase catalytic activity of Fe_7S_8 nanowires templated from $[Fe_{16}S_{20}]$/diethylenetriamine hybrid nanowires. ChemPlusChem. 2013;**78**:723-772. DOI: 10.1002/cplu.201300075

[95] Dutta AK, Das S, Samanta S, Samanta PK, Adhikary B, Biswas P. CuS nanoparticles as a mimic peroxidase for colorimetric estimation of human blood glucose level. Talanta. 2013;**107**:361-367. DOI: 10.1016/j.talanta.2013.01.032

[96] Dalui A, Pradhan B, Thupakula U, Khan AH, Kumar GS, Ghosh T, et al. Insight into the mechanism revealing the peroxidase mimetic catalytic activity of quaternary CuZnFeS nanocrystals: Colorimetric biosensing of hydrogen peroxide and glucose. Nanoscale. 2015;**7**:9062-9074. DOI: 10.1039/c5nr01728a

[97] Maji SK, Dutta AK, Srivastava DN, Paul P, Mondal A, Adhikary B.

Peroxidase-like behavior, amperometric biosensing of hydrogen peroxide and photocatalytic activity by cadmium sulfide nanoparticles. Journal of Molecular Catalysis A: Chemical. 2012;**358**:1-9. DOI: 10.1016/j.molcata.2012.03.007

[98] Xu J, Cai R, Zhang Y, Mu X. Molybdenum disulfide-based materials with enzyme-like characteristics for biological applications. Colloids and Surfaces B: Biointerfaces. 2021;**200**:111575. DOI: 10.1016/j.colsurfb.2021.111575

[99] Lin T, Zhong L, Song Z, Guo L, Wu H, Guo Q, et al. Visual detection of blood glucose based on peroxidase-like activity of WS_2 nanosheets. Biosensors and Bioelectronics. 2014;**62**:302-307. DOI: 10.1016/j.bios.2014.07.001

[100] Dutta AK, Maji SK, Srivastava DN, Mondal A, Biswas P, Paul P, et al. Synthesis of FeS and FeSe nanoparticles from a single source precursor: A study of their photocatalytic activity, peroxidase-like behavior, and electrochemical sensing of H_2O_2. ACS Applied Materials and Interfaces. 2012;**4**:1919-1927. DOI: 10.1021/am300408r

[101] Wang Y, Zhang D, Xiang Z. Synthesis of α-MnSe crystal as a robust peroxidase mimic. Materials Research Bulletin. 2015;**67**:152-157. DOI: 10.1016/j.materresbull.2015.03.018

[102] Wu X, Chen T, Wang J, Yang G. Few-layered $MoSe_2$ nanosheets as an efficient peroxidase nanozyme for highly sensitive colorimetric detection of H_2O_2 and xanthine. Journal of Materials Chemistry B. 2017;**6**(1):105-111. DOI: 10.1039/c7tb02434g

[103] Khagar P, Pratap UR, Zodape SP, Wankhade AV. Self-assembled CoSe2 microspheres with intrinsic peroxidase mimicking activity for efficient degradation of variety of dyes. ChemistrySelect. 2021;**6**(20):5043-5051. DOI: 10.1002/slct.202101496

[104] Beinert H, Holm RH, Münck E. Iron-sulfur clusters: Nature's modular, multipurpose structure. Science. 1997;**277**:653-659. DOI: 10.1126/science.277.5326.653

[105] Imlay JA. Iron-sulphurclusters and the problem with oxygen. Molecular Microbiology. 2006;**59**(4):1073-1082. DOI: 10.1111/j.1365-2958.2006.05028.x

[106] Crack JC, Le Brun NE. Biological iron-sulfur clusters: Mechanistic insights from mass spectrometry. Coordination Chemistry Reviews. 2021;**446**:214171. DOI: 10.1016/j.ccr.2021.214171

[107] Poole LB. Flavin-dependent alkyl hydroperoxide reductase from salmonella typhimurium. 2. Cystine disulfides involved in catalysis of peroxide reduction. Biochemistry. 1996;**35**(1): 65-75. DOI: 10.1021/bi951888k

[108] Hall A, Nelson K, Poole LB, Karplus PA. Structure-based insights into the catalytic power and conformational dexterity of peroxiredoxins. Antioxidants and Redox Signaling. 2011;**15**:795-815. DOI: 10.1089/ars.2010.3624

[109] Cairns-Smith AG. Seven Clues to the Origin of Life - a Scientific Detective Story. Cambridge, UK: Cambridge University Press; 1990

[110] Camprubi E, de Leeuw JW, House CH, Raulin F, Russell MJ, Spang A, et al. The Emergency of Life. Space Science Reviews. 2019;**215**:56. DOI: 10.1007/s11214-019-0624-8

[111] Gilbert W. Origin of life: The RNA world. Nature. 1986;**319**:618. DOI: 10.1038/319618a0

[112] Higgs P, Lehman N. The RNA world: Molecular cooperation at the origins of

life. Nature Reviews Genetics. 2015;**16**:7-17. DOI: 10.1038/nrg3841

[113] Ferus M, Adam V, Cassone G, et al. Ariel – A window to the origin of life on early earth? Experimental Astronomy. 2020. DOI: 10.1007/s10686-020-09681-w

[114] Duval S, Branscomb E, Trolard F, Bourrié G, Grauby O, Heresanu V, et al. On the why's and how's of clay minerals' importance in life's emergence. Applied Clay Science. 2020;**195**:105737. DOI: 10.1016/j.clay.2020.105737

[115] Russell MJ. The "water problem"(sic), the illusory pond and life's submarine emergence-a review. Life (Basel). 2021;**11**(5):429. DOI: 10.3390/life11050429

[116] Jambor JL, Dutrizac JE. Occurrence and constitution of natural and synthetic ferrihydrite, a widespread iron oxyhydroxide. Chemistry Review. 1998;**98**(7):2549-2586. DOI: 10.1021/cr970105t

[117] Guo A, Barnard S. Naturally occurring iron oxide nanoparticles: Morphology, surface chemistry and environmental stability. Journal of Materials Chemistry A. 2013;**1**:27-42. DOI: 10.1039/C2TA00523A

[118] Bjorn VDH, Roychoudhury A, Myneni S. Iron-rich nanoparticles in natural aquatic environments. Minerals. 2019;**9**:287. DOI: 10.3390/min9050287

[119] Wu J, Boyle E, Sunda W, Wen LS. Soluble and colloidal iron in the oligotrophic North Atlantic and North Pacific. Science. 2001;**293**(5531):847-849. DOI: 10.1126/science.1059251

[120] Fortin D, Langley S. Formation and occurrence of biogenic iron-rich minerals. Earth-Science Reviews. 2005;**72**:1-19. DOI: 10.1111/j.1472-4669.2007.00131.x

[121] Zhang XL, Wu GJ, Zhang CL, et al. What is the real role of iron oxides in the optical properties of dust aerosols. Atmosheric Chemistry and Physics. 2015;**15**(21):12159-12177. DOI: 10.5194/ACP-15-12159-2015

[122] Berner RA. Phosphate removal from sea water by adsorption on volcanogenic ferric oxides. Earth and Planetary Science Letters. 1973;**18**(1): 77-86. DOI: 10.1016/0012-821X(73)90037-X

[123] Delaney ML. Phosphorus accumulation in marine sediments and the oceanic phosphorus. Global Biogeochemical Cycles. 1998;**12**(4): 563-572. DOI: 10.1029/98GB02263

[124] Benitez-Nelson CR. The biogeochemical cycling of phosphorus in marine systems. Earth-Science Reviews. 2000;**51**(1-4):109-135. DOI: 10.1016/S0012-8252(00)00018-0

[125] Paytan A, McLaughlin K. The oceanic phosphorus cycle. Chemical Reviews. 2007;**107**(2):563-576. DOI: 10.1021/cr0503613

[126] Tagliabue A, Bowie AR, Boyd PW, Buck KN, Johnson KS, Saito MA. The integral role of iron in ocean biogeochemistry. Nature. 2017;**543**(7643):51-59. DOI: 10.1038/nature21058

[127] Walton CR, Shorttle O, Jenner FE, et al. Phosphorus mineral evolution and prebiotic chemistry: From minerals to microbes. Earth-Science Reviews. 2021;**221**:103806. DOI: 10.1016/j.earscirev.2021.103806

[128] McLaren TI, Smernik RJ, McLaughlin MJ, McBeath TM, Kirby JK, Simpson RJ, et al. Complex forms of soil organic phosphorus-a major component of soil phosphorus. Environment Science

& Technology. 2015;**49**(22):13238-13245. DOI: 10.1021/acs.est.5b02948

[129] Huang LM, Jia XX, Zhang GL, et al. Soil organic phosphorus transformation during ecosystem development: A review. Plant & Soil. 2017;**417**:17-42. DOI: 10.1007/s11104-017-3240-y

[130] Zhang Q, Wang Y, Zhang C, et al. A review of organophosphate esters in soil: Implications for the potential source, transfer, and transformation mechanism. Environmental Research. 2022;**204**:112122. DOI: 10.1016/j.envres.2021.112122

[131] Ged EC, Boyer TH. Molecular weight distribution of phosphorus fraction of aquatic dissolved organic matter. Chemosphere. 2013;**91**:921-927. DOI: 10.1016/j.chemosphere.2013.01.113

[132] Douglas WB, Pellechia P, Chambers LR, et al. Isolation and molecular characterization of dissolved organic phosphorus using electrodialysis-reverse osmosis and solution 31P-NMR. Limnology and Oceanography Methods. 2017;**15**:436-452. DOI: 10.1002/lom3.10171

[133] Feng W, Zhu Y, Wu F, et al. Characterization of phosphorus forms in lake macrophytes and algae by solution (31) P nuclear magnetic resonance spectroscopy. Environmental Science and Pollution Research International. 2016;**23**(8):7288-7297. DOI: 10.1007/s11356-015-5913-5

[134] Bell DW, Pellechia PJ, Ingall ED, Benitez-Nelson CR. Resolving marine dissolved organic phosphorus (DOP) composition in a coastal estuary. Limnology and Oceanography. 2020; **65**(11):2787-2799. DOI: 10.1002/lno.11552

[135] Baken S, Regelink IC, Comans RNJ, Smolders E, Koopmans GF. Iron-rich colloids as carriers of phosphorus in streams: A field-flow fractionation study. Water Research. 2016;**99**:83-90. DOI: 10.1016/j.watres.2016.04.060

[136] Gottselig N, Amelung W, Kirchner JW, Bol R, Eugster W, et al. Elemental composition of natural nanoparticles and fine colloids in European forest stream waters and their role as phosphorus carriers. Global Biogeochem Cycles. 2017;**31**:1592-1607. DOI: 10.1002/2017GB005657

[137] Missong A, Holzmann S, Bol R, Nischwitz V, Puhlmann H, Wilpert KV, et al. Leaching of natural colloids from forest topsoils and their relevance for phosphorus mobility. Science of the Total Environment. 2018;**634**:305-315. DOI: 10.1016/j.scitotenv.2018.03.265

[138] Li F, Zhang Q, Klumpp E, Bol R, Nischwitz V, Ge Z, et al. Organic carbon linkage with soil colloidal phosphorus at regional and field scales: Insights from size fractionation of fine particles. Environmental Science & Technology. 2021;**55**(9):5815-5825. DOI: 10.1021/acs.est.0c07709

[139] Fresne M, Jordan P, Fenton O, Mellander PE, Daly K. Soil chemical and fertilizer influences on soluble and medium-sized colloidal phosphorus in agricultural soils. Science of the Total Environment. 2021;**754**:142112. DOI: 10.1016/j.scitotenv.2020.142112

[140] Chekli L, Phuntsho S, Roy M, Shon HK. Characterisation of Fe-oxide nanoparticles coated with humic acid and Suwannee River natural organic matter. Science of the Total Environment. 2013;**461-462**:19-27. DOI: 10.1016/j.scitotenv.2013.04.083

[141] Vindedahl AM, Strehlau JH, Arnold WA. Organic matter and iron oxide nanoparticles: Aggregation, interactions, and reactivity (review). Environmental

Science: Nano. 2016;**3**(3):494-505. DOI: 10.1039/C5EN00215J

[142] Bao Y, Bolan NS, Lai J, et al. Interactions between organic matter and Fe (hydr)oxides and their influences on immobilization and remobilization of metal(loid)s: A review. Critical Reviews in Environmental Science and Technology. 2021:1-22. DOI: 10.1080/10643389.2021.1974766

[143] Wang YX, Zhao MT, Ping JF, Chen B, Cao XH, Huang Y, et al. Bioinspired design of ultrathin 2D bimetallic metal-organic-framework nanosheets used as biomimeticenzymes. Advanced Materials. 2016;**28**:4149-4155. DOI: 10.1002/adma.201600108

[144] Fan K, Wang H, Xi J, et al. Optimization of Fe_3O_4 nanozyme activity via single amino acid modification mimicking an enzyme active site. Chemical Communications. 2017;**53**(2):424-427. DOI: 10.1039/c6cc08542c

[145] Zhang Z, Li Y, Zhang X, Liu J. Molecularly imprinted nanozymes with faster catalytic activity and better specificity. Nanoscale. 2019;**11**(11): 4854-4863. DOI: 10.1039/c8nr09816f

[146] Ling PH, Cheng S, Chen N, Qian CH, Gao F. Nanozyme-modified metal-organic frameworks with multienzymes activity as biomimetic catalysts and electrocatalytic interfaces. ACS Applied Materials & Interfaces. 2020;**12**:17185-17192. DOI: 10.1021/acsami.9b23147

[147] Liu Q, Zhang A, Wang R, Zhang Q, Cui D. A review on metal- and metal oxide-based nanozymes: Properties, mechanisms, and applications. Nano-Micro Letters. 2021;**13**(1):1-53. DOI: 10.1007/s40820-021-00674-8

[148] Redfield AC. The biological control of chemical factors in the environment. American Scientist. 1958;**46**(3):205-221

[149] Nature Geoscience Editorial. Eighty years of Redfield. Nature Geoscience. 2014;**7**:849. DOI: 10.1038/ngeo2319

[150] Lenton TM, Watson AJ. Redfield revisited: 1. Regulation of nitrate, phosphate, and oxygen in the ocean. Global Biogeochemical Cycles. 2000;**14**(1):225-248. DOI: 10.1029/1999GB900065

[151] Geider RJ, La Roche J. Redfield revisited: Variability of C:N:P in marine microalgae and its biochemical basis. European Journal of Phycology. 2002;**37**(1):1-17. DOI: 10.1017/S0967026201003456

[152] Sañudo-Wilhelmy S, Tovar-Sanchez A, Fu FX, et al. The impact of surface-adsorbed phosphorus on phytoplankton Redfield stoichiometry. Nature. 2004;**432**:897-901. DOI: 10.1038/nature03125

[153] Duhamel S, Diaz JM, Adams JC, et al. Phosphorus as an integral component of global marine biogeochemistry. Nature Geoscience. 2021;**14**:359-368. DOI: 10.1038/s41561-021-00755-8

[154] Le Quéré C, Harrison SP, Prentice IC, et al. Ecosystem dynamics based on plankton functional types for global ocean biogeochemistry models. Global Change Biology. 2005;**11**:2016-2040. DOI: 10.1111/j.1365-2486.01004.x

[155] Moore C, Mills M, Arrigo K, et al. Processes and patterns of oceanic nutrient limitation. Nature Geoscience. 2013;**6**:701-710. DOI: 10.1038/ngeo1765

[156] Matsumoto K, Tanioka T, Rickaby R. Linkages between dynamic phytoplankton C:N:P and the ocean carbon cycle under climate change. Oceanography. 2020;**33**(2):44-52. DOI: 10.5670/oceanog.2020.203

[157] Martin JH. Glacial-interglacial CO2 change: The Iron Hypothesis. Paleoceanography. 1990;**5**(1):1-13. DOI: 10.1029/PA005i001p0000

[158] Boyd PW, Watson AJ, Law CS, et al. A mesoscale phytoplankton bloom in the polar Southern Ocean stimulated by iron fertilization. Nature. 2000;**407**(6805): 695-702. DOI: 10.1038/35037500

[159] Boyd PW, Jickells T, Law CS, et al. Mesoscale iron enrichment experiments 1993-2005: Synthesis and future directions. Science. 2007;**315**(5812): 612-617. DOI: 10.1126/science.1131669

[160] Buesseler KO, Doney SC, Karl DM, et al. Ocean iron fertilization--Moving forward in a sea of uncertainty. Science. 2008;**319**(5860):162. DOI: 10.1126/science.1154305

[161] UNEP. The Emissions Gap Report. 2016. Available from: https://www.unep.org/resources/publication/emissions-gap-report-2016-un-environment-synthesis-report

[162] Emerson D. Biogenic iron dust: A novel approach to ocean iron fertilization as a means of large scale removal of carbon dioxide from the atmosphere. Frontiers in Marine Science. 2019;**6**:22. DOI: 10.3389/fmars.2019.00022

[163] Scott-Buechler CM, Greene CH. Chapter 6 - Role of the ocean in climate stabilization. In: Magalhães Pires JC, Da Cunha Gonçalves AL, editors. Bioenergy with Carbon Capture and Storage: Using Natural Resources for Sustainable Development. Academic Press; 2019. pp. 109-130. DOI: 10.1016/B978-0-12-816229-3.00006-5

[164] Braunschweig J, Bosch J, Meckenstock RU. Iron oxide nanoparticles in geomicrobiology: From biogeochemistry to bioremediation. New Biotechnology. 2013;**30**(6):793-802. DOI: 10.1016/j.nbt.2013.03.008

[165] Baumgartner J, Faivre D. Iron solubility, colloids and their impact on iron (oxyhydr)oxide formation from solution. Earth-Science Reviews. 2015;**150**:520-530. DOI: 10.1016/j.earscirev.2015.09.003

[166] Claudio C, Di Iorio E, Liu Q, Jiang Z, Barrón V. Iron oxide nanoparticles in soils: Environmental and agronomic importance (review). Journal of Nanoscience and Nanotechnology. 2017;**17**(7):4449-4460. DOI: 10.1166/jnn.2017.14197

[167] Wu J, Wang X, Wang Q, Lou Z, Li S, Zhu Y, et al. Nanomaterials with enzyme-like characteristics (nanozymes): Next-generation artificial enzymes (II). Chemical Society Reviews. 2019;**48**(4): 1004-1076. DOI: 10.1039/c8cs00457a

[168] Malhotra N, Lee JS, Liman R, Ruallo J, Villaflores OB, Ger TR, et al. Potential toxicity of iron oxide magnetic nanoparticles: A review. Molecules (Basel, Switzerland). 2020;**25**(14):3159. DOI: 10.3390/molecules25143159

Section 2

Synthesis

Chapter 2

New Approaches in Synthesis and Characterization Methods of Iron Oxide Nanoparticles

Cristina Chircov and Bogdan Stefan Vasile

Abstract

Recent years have witnessed an extensive application of iron oxide nanoparticles within a wide variety of fields, including drug delivery, hyperthermia, biosensing, theranostics, and cell and molecular separation. Consequently, synthesis and characterization methods have continuously evolved to provide the possibility for controlling the physico-chemical and biological properties of the nanoparticles to better suit the envisaged applications. In this manner, this chapter aims to provide an extensive overview of the most recent progress made within the processes of iron oxide nanoparticle synthesis and characterization. Thus, the chapter will focus on novel and advanced approaches reported in the literature for obtaining standardized nanoparticles with controllable properties and effects. Specifically, it will emphasize the most recent progress made within the microwave-assisted, microfluidics, and green synthesis methods, as they have shown higher capacities of controlling the outcome nanoparticle properties.

Keywords: iron oxide nanoparticles, synthesis processes, green synthesis processes, characterization techniques, physico-chemical properties

1. Introduction

Iron oxides are transition metal oxides ubiquitously found in nature, having many implications in various biological and geological processes [1–4]. They occur naturally as aggregates, mineral nanoparticles, or nanostructured coatings onto other soil grains [5], being an essential biogeochemically-active component of the Earth ecosystem [6]. Moreover, iron oxides are formed in a variety of polymorphs with different stoichiometric and crystalline structures [2, 3, 7], including oxides, e.g., wüstite or ferrous oxide (FeO), magnetite (Fe_3O_4 or $FeO{\cdot}Fe_2O_3$), maghemite (γ-Fe_2O_3), ε-Fe_2O_3, hematite (α-Fe2O3), and β-Fe_2O_3, hydroxides, e.g., iron(III) hydroxide (bernalite) and iron(II) hydroxide, and oxyhydroxydes, e.g., goethite, feroxyhyte, akaganeite, and lepidocrocite [2, 3, 7–13]. Among them, magnetite, maghemite, and hematite are crystalline and most commonly used in biomedical and pharmaceutical applications, while other forms, such as goethite, are amorphous and occur at high pressure and temperature conditions [2, 3, 7, 14]. Furthermore, iron oxides can also be categorized based on their electrical properties into insulative, i.e., ferrous oxide, conductive, i.e., magnetite, and semiconductive, i.e., hematite and goethite [1].

By contrast, iron oxides can be processed into nanoparticles with magnetic properties, which further allow for their manipulation by external magnetic fields [7, 15]. Among them, magnetite nanoparticles are, by far, the most intensively studied, as they have demonstrated considerable potential in a myriad of applications, including drug delivery, magnetofection, hyperthermia, photoablation therapy, magnetic resonance imaging as contrast enhancement agents, theranostics, biosensing, bioanalysis through biological labeling, tracking, and detection, bioseparation, antimicrobial therapies, tissue engineering and regeneration, wound healing, catalysis, nanorobots, ferrofluids, microelectronics and ultrahigh density magnetic storage media, magnetic paints, pollutant removal sorbents, and batteries [7, 16–22]. Additionally, recent studies have shown an intrinsic peroxidase activity of iron oxide nanoparticles, which could be further exploited in applications such as biocatalysis, wastewater treatment, detection tools, magnetic enzyme-linked immunosorbent assay kits, or artificial enzymes [23–29].

Evidently, each of the previously mentioned applications require specific nanoparticle properties [30]. In this context, it has been confirmed that the physico-chemical properties of magnetite nanoparticles, namely, size, shape, stability, crystal structure and crystallinity, chemical composition, and surface area, energy, and roughness, significantly determine their magnetic properties and, consequently, their biological behavior, drug concentration, toxicity, and efficacy [19, 31–33]. Moreover, studies have shown that the synthesis route of magnetite nanoparticles greatly impacts their physico-chemical properties, thus highlighting the necessity to improve synthesis performance by enhancing standardization, automation, monitoring, and mass production [17, 19, 33].

Presently, the most ubiquitous synthesis method for magnetite nanoparticles is the co-precipitation of ferrous and ferric ions through the addition of an alkaline solution [34]. Although it is a simple and cost-efficient method characterized by considerably high productivity [32], its reproducibility is still limited due to the presence of the intermediate phases within the final product. Additionally, it does not allow for the precise control of nanoparticle size and shape, which further leads to significant variations in the physico-chemical properties of the final product [32, 34–36]. Therefore, there is a fundamental need for the exploration of novel synthesis processes that could further ensure optimal, controllable, and scalable properties. In this context, iron oxide nanoparticles obtained from natural, green sources are continuously gaining the interest of the scientific community as they provide a potential alternative to overcome the limitations of conventional nanoparticles [37]. Consequently, characterization techniques should also be advanced to ensure a reliable assessment of nanoparticle properties. In this manner, the variety of magnetic nanoparticle applications, ranging from biomedicine and pharmaceutical industry to data storage, could greatly benefit from such improvements.

Therefore, this chapter aims to provide an updated overview of the most recent developments within the field of magnetite nanoparticle synthesis and processing, as well as the most advanced characterization techniques utilized for evaluating their properties and potential.

2. Novel iron oxide nanoparticles synthesis methods

There are two well-established approaches involved in the synthesis of nanoparticles, namely, top-down and bottom-up approaches. Generally, top-down methods involve the crushing, breaking, or fractioning of bulk materials into smaller parts to produce nanoparticles through mechanical action [38, 39]. Such methods include mechanical crushing, milling, or grinding, laser ablation, sputtering, etching, or electron beam deposition, offering an alternative eco-friendly route despite the required high time and power consumption [40–42]. By contrast, bottom-up approaches are based on chemical reactions among specific atoms, ions, or molecules necessary for the formation of nanoparticles. Considering these principles, synthesis routes can be further divided based on the nature of the involved process into physical, that can be associated to the top-down methods, chemical (e.g., co-precipitation, sol-gel, thermal decomposition, emulsion and microemulsion, hydrothermal, and microwave-assisted methods), and biological (which utilize plants or microorganisms for the generation of nanoparticles), the latter two being attributed to bottom-up approaches (**Figure 1**) [38, 39, 43, 44].

Figure 2 depicts a comparison between the most commonly used methods for the synthesis of magnetite nanoparticles. It can be observed that chemical methods comprise the majority of the investigated routes, the co-precipitation method accounting for the highest percentage. Specifically, the co-precipitation process involves two possible pathways, either partial oxidation of iron(II) salts or the aging of a stoichiometric mixture of iron(III) and iron(II) salts through the addition of an alkaline solution that leads to nucleation and growth mechanisms and finally to the generation of Fe_3O_4 nanoparticles. The chemical reaction principle involved in the production of magnetite nanoparticles is shown in Eq. (1):

$$Fe^{2+} + 2Fe^{3+} + 8OH^{-} \leftrightarrow Fe(OH)_2 + 2Fe(OH)_3 \rightarrow Fe_3O_4 + 4H_2O. \quad (1)$$

Although it is the easiest to implement, time-efficient, and safe method, involving limited use of harmful solvents, the co-precipitation process is considerably disadvantageous in terms of reproducibility and possibility to control the outcome properties of the obtained nanoparticles [46, 47].

Thus, there is a fundamental need for the investigation of novel synthesis routes that could improve the features of magnetite nanoparticles. In this context, recent years have witnessed a shift toward the implementation of previously non-conventional methods that could potentially provide a plethora of alternatives in terms of modulating physico-chemical properties. Thus, the following sections will describe the most recent advancements within the production of magnetite nanoparticles through microwave-assisted, microfluidic, and green synthesis methods.

2.1 Microwave-assisted method

Owing to its numerous advantages, microwave-assisted synthesis has become a particularly attractive method for various synthetic chemistry reactions. Specifically, this technique has provided the means for the easy production of nanoparticles in a considerably time- and cost-efficient manner, with reduced energy consumption and increased environmental friendliness [48–51] through the use of 50% less power than electric furnaces with similar capacities [52]. Besides the associated economic aspects, the microwave-assisted method has received increased scientific interest due to the possibility of tuning the parameters to obtain the desired size and shape of magnetite nanoparticles with significantly narrow distributions and high reproducibility, phase purity, and yield [49–51]. This is possible due to the characteristic uniform heating and nucleation, rapid kinetics and crystallization, and phase selectivity [50, 51].

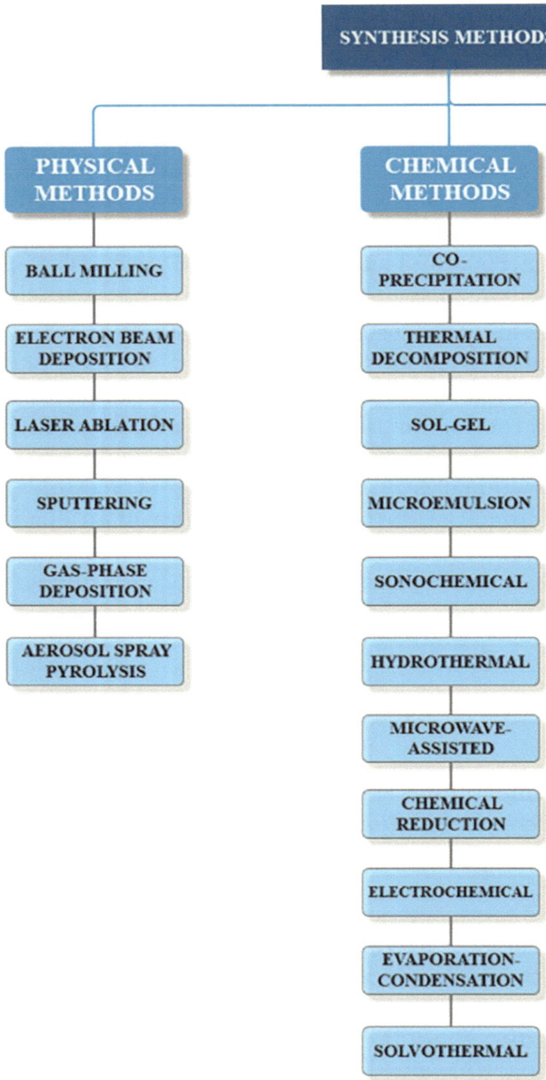

Figure 1.
The main types of magnetite nanoparticle synthesis methods. Reprinted from an open-access source [43].

The basic principle involved in this method is based on the activation and subsequent alignment of dipoles (i.e., mechanism of dipolar polarization) and/or ions (i.e., mechanism of ionic conduction) present within a material through the interactions with microwave electromagnetic radiations. Consequently, internal heating will occur in a highly homogenous manner, thus, leading to a rapid temperature rise that is responsible for reducing the reaction time and the necessary energy [48–52]. In the case of magnetite nanoparticles, it has been demonstrated that the microwave-assisted method offers the possibility to control their magnetic properties by adjusting the experimental parameters [49].

Generally, the microwave-assisted method is combined with other synthesis processes, such as co-precipitation. Thus, the synthesis involves the co-precipitation

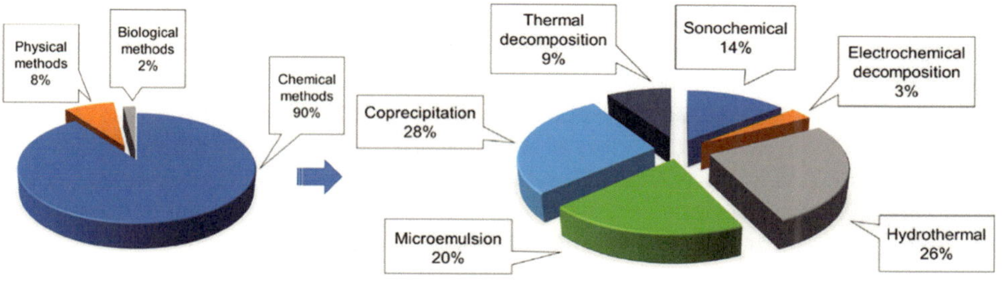

Figure 2.
The prevalence of the most commonly utilized magnetite nanoparticle synthesis methods. Reprinted from an open-access source [45].

of iron oxide nanoparticles through the classical method, followed by the microwave treatment that enables the control over the properties of the nanoparticles. Several studies have demonstrated the possibility to obtain monodisperse iron oxide nanoparticles with well-controlled sizes and high crystallinity, saturation magnetizations, and stability, as compared to the co-precipitation counterparts [53–56]. Additionally, the possibility of developing uniform polyethylene glycol [53], humate polyanion [54], and silica [55] coatings was also demonstrated.

2.2 Microfluidic approaches

Microfluidics is a relatively new field that has brought together fluid dynamics, chemistry, and material science principles for allowing the precise and accurate manipulation of small fluid volumes within microchannels [57, 58]. In this context, microfluidic technology-based methods for the synthesis of nanomaterials have emerged as an alternative to conventional routes that could provide possible solutions to the currently existing limitations. Specifically, microfluidic devices represent synthesis platforms with outstanding features for the fabrication of nanoparticles, including small capillary dimensions and consequent large surface/volume ratios and reduced reagent volume use, rapid and uniform mass and heat transfer, ease of automation, reduced residence time, and precise control of mixing [38, 57–59]. In this manner, by increasing the control of the implicated reaction parameters (e.g., device geometry, flow rate, reagent concentration, reaction time, temperature) [59, 60], nanoparticles with superior uniformity, stability, and encapsulation efficiency and narrow particle size distributions can be obtained in a highly reproducible and controllable manner [38, 57–60].

Based on their geometry, microfluidic devices can be classified into tubular reactors, which generally involve circular channels that are either in-house produced or purchased (most common commercially available reactors have T and Y type junctions), and chip reactors, which involve more complex geometries and are usually in-house manufactured using various fabrication techniques [38]. The working principle of microfluidic approaches for the synthesis of nanoparticles resides on the movement of fluids within microchannels and microchambers with unique geometries to integrate the preparation, reaction, and separation steps. In this context, there are two main types of microfluidic reactors that involve different synthesis strategies, namely, single-phase or continuous-flow microfluidics and multi-phase or droplet-based microfluidics, which can be further divided according to the carrier fluid into gas–liquid and liquid–liquid segmented flows [38, 59].

Microfluidics is currently evolving as a promising alternative for the synthesis of magnetite nanoparticles with controlled size, shape, and surface chemistry that can be modulated according to the application requirements [57, 58]. Since it involves a relatively simple reaction, it can be obtained through both types of synthesis strategies. Continuous-flow microreactors that contain one inlet for the iron precursor solution and one inlet for the alkaline solution will ensure the formation of the nanoparticles at the interface between the two fluid layers if the pH value is high enough for nucleation. While this is usually the preferred route owing to its increased homogeneity and versatility, some applications require faster interactions. Therefore, the multiphase microfluidics involving cross-flow designs are receiving increasing attention. In this approach, the channels containing the precursor solutions, i.e., the dispersed phase, will intersect the channels containing the alkaline solution, i.e., the continuous phase, where the nanoparticles will form and be further transported within the continuous phase [38, 60]. Furthermore, the microfluidic platforms used for the synthesis of magnetite nanoparticles can be made of various materials, such as glass, metals, silicon, or polymers, that must be resistant to the fluids introduced within the microchannels [38].

Although the number of studies is still limited, the results are promising, thus paving the way toward the future of nanoparticle synthesis. In this context, the synthesis of magnetite nanoparticles was investigated through the use of a single-flow polydimethylsiloxane microfluidic reactor [61] and a T-junction polymethylmethacrylate microchip fabricated by a laser cutting machine [57]. Furthermore, another study fabricated magnetite nanoparticles using a 3D flow microfluidic device focused by two basic sheath streams, that were subjected to a postsynthesis surface functionalization step outside the microreactor [62]. Moreover, other studies demonstrated the possibility of developing *in situ* chitosan-coated magnetite nanoparticles using two types of microchip configurations fabricated through 3D printing [63] and by the soft lithography process [64].

2.3 Green synthesis methods

The merge between nanotechnology and biology has led to the rise of a new and highly advanced field of nanomaterial synthesis using living microorganisms of both prokaryotic and eukaryotic origins, such as algae, bacteria, fungi, yeasts, viruses, and plants [65]. Within this framework, the synthesis of nanoparticles via green technologies utilizing microorganisms and plant extracts is continuously emerging as a safe, cost-efficient, renewable, and environmentally friendly alternative [65–68] which does not implicate complex protocols [69] or the use of intermediary base groups [70]. Additionally, green synthesis methods lead to the formation of nanoparticles with higher stability as they do not involve the use of chemicals that increase particle reactivity, enhanced biocompatibility, non-toxicity, and antimicrobial and anticancer properties [66–69]. Such methods are possible due to the resistance mechanisms developed by microorganisms and plants to endure the highly toxic environments generated by high metal concentrations. Specifically, the intrinsic chemical processes of these living entities can remodel inorganic metal ions into nanoparticles to reduce or eliminate the toxic effects. There are two main processes involved in the biogenic synthesis of nanoparticles, namely, through bioreduction, i.e., the reduction of metal ions by intrinsic biological processes, and biosorption, involving the assimilation of metal ions within the cell wall and the consequent formation of stable nanoparticulate structures through the assembly with the present macromolecules [65].

Generally, plant-based synthesis of nanoparticles is more advantageous in terms of higher kinetics, increased reduction and stabilization yield, and easier large-scale production [66, 68, 70]. The plant-mediated formation of nanoparticles can occur intracellularly or inside the plant, through the presence of specific biomolecules (e.g., aldehydes, ketones, flavones, phenols, amino acids, proteins, polysaccharides, tannins, terpenoids, saponins, vitamins), extracellularly, using plant extracts, or through individual phytochemicals. The mechanism involves the linkage between the atmospheric or phytochemical-generated oxygen that reduces the metal ions, followed by the electrostatic interactions between the newly formed metal oxides that will lead to the formation of the nanoparticles. The nature of the phytochemicals is responsible for their size, shape, stability, and reactivity variations [67, 68, 71]. Based on their produced phytochemicals, various plant parts have been investigated, including root, leaf, flower, petal, fruit, stem, peel, or seed [70].

Although green synthesis methods have been mostly applied for obtaining silver, gold, and copper nanoparticles, the synthesis of iron oxide nanoparticles through the use of plants or microorganisms has become an intensively studied field owing to the biocompatible, non-toxic, and stable nature of the final products [72]. Thus, there are many protocols available in the literature for the green synthesis of iron oxide nanoparticles, which generally follow a similar methodology. Briefly, the procedure begins with the starting material preparation and extraction, by collecting, washing, drying, weighing, grinding into a fine powder, boiling in water or methanol/ethanol under continuous stirring, centrifugation, and filtration. Subsequently, the extract is mixed with the precursor salt solutions, such as $FeSO_4{\cdot}7H_2O$, $FeCl_3{\cdot}6H_2O$, $(FeNO_3)_3{\cdot}9H_2O$, $FeSO_4$, $FeCl_3$, $FeCl_2{\cdot}4H_2O$, $FeCl_2$, or $FeSO_4{\cdot}5H_2O$, of varying molarities. Finally, the mixture is heated and vigorously stirred until the color of the solution changes and intensifies according to the type of iron salts utilized. The obtained iron oxide nanoparticle pellets are further washed and dried [73].

The available literature studies reported the synthesis of iron oxide nanoparticles using Bauhinia tomentosa [74], pomegranate seeds [75], Hibiscus rosa-sinensis [76], Mimosa pudica root [77], Carica papaya leaf extract [78], Cymbopogon citratus [79], Ficus carica leaf extract [80], and Platanus orientalis leaf extract [81].

3. Advanced iron oxide nanoparticles characterization techniques

As previously emphasized, the physico-chemical properties of iron oxide nanoparticles often dictate their applications [82]. Since the characterization of nanoparticles is significantly challenging due to the increased interdisciplinarity of the field, it is fundamentally important to characterize nanoparticles to the maximum extent to ensure a more rapid implementation in commercial applications [83]. Generally, the physicochemical properties of iron oxide nanoparticles are evaluated through a variety of different techniques, depending on the parameter that must be determined [35, 73, 82]. **Table 1** depicts the most important characteristics of iron oxide nanoparticles and the suitable characterization techniques for determining them.

The most important parameter to be evaluated is the size and consequently the size distribution of the nanoparticles, as it can affect other properties and determine the behavior of the final product within the envisaged application [35, 83]. Although size measurements within the macroscale might appear trivial, size determinations within the nanoregime might lead to different interpretations depending on the

Nanoparticle property	Characterization techniques
Size	TEM, XRD, DLS, NTA, HRTEM, SAXS, SEM, AFM
Size distribution	DLS, NTA, SAXS
Shape	TEM, HRTEM, SEM, AFM, STEM
Crystal structure	XRD, HRTEM, SAED, STEM
Elemental/chemical composition	XRD, SEM-EDX, ICP-MS, XPS, EELS
Surface area/specific surface area	BET analysis, NMR
Surface charge	Zeta potential
Magnetic properties	VSM, SQUID, Mössbauer spectroscopy, MFM

TEM—transmission electron microscopy, XRD—X-ray diffraction, DLS—dynamic light scattering, NTA—nanoparticle tracking analysis, HRTEM—high-resolution TEM, SAXS—small-angle X-ray scattering, SEM—scanning electron microscopy, AFM—atomic force microscopy, STEM—scanning transmission electron microscope, SAED—selected area electron diffraction, EDX—energy-dispersive X-ray spectroscopy, ICP-MS—inductively coupled plasma mass spectrometry, XPS—X-ray photoelectron spectroscopy, EELS—electron energy loss spectroscopy, BET—Brunauer–Emmett–Teller, NMR—nuclear magnetic resonance, VSM—vibrating sample magnetometry, SQUID—superconducting quantum interference device, MFM—magnetic force microscopy.

Table 1.
The characteristics of iron oxide nanoparticles and the associated characterization techniques. Adapted from an open-access source [35, 73, 83].

characterization technique employed. When referring to nanoparticles, size can be correlated to the atomic structure-defined physical dimension, the diffusion/sedimentation-dependent effective size of the nanoparticle within a matrix or solvent, or the effective size weighted by the mass/electron distribution [84, 85]. Furthermore, size distribution represents an estimation of the quality of the synthesis process, as the general aim is to obtain close to monodisperse nanoparticles [82–84].

The shape also plays a fundamental role upon the behavior of iron oxide nanoparticles, as it can further lead to toxic effects due to cell harming. Therefore, the employed synthesis routes must allow for the control of nanoparticle shape as a crucial parameter [86, 87]. Commonly, electron microscopy techniques are utilized for the precise evaluation of the morphology and consequently the shape of the nanoparticles [83, 88].

Moreover, crystal structure and chemical composition also represent essential characterization steps in the process of iron oxide nanoparticle development [83]. Although X-ray diffraction represents the most common technique for crystal structure, crystallinity, and phases evaluation [89], studies have shown that in the case of nanoparticles with sizes below 5 nm, the diffractogram patterns are influenced [90]. Thus, other, more reliable methods should be developed. Selected area electron diffraction represents an alternative that better depicts the crystal structure of nanoparticles [84]. Chemical and elemental composition determination provides an estimation of the purity of the nanoparticles. Additionally, the chemical composition is a key parameter that can influence the electrochemical activity of the nanoparticles [91]. Moreover, another interesting characterization possibility involves the precise distinction between iron(II) and iron(III) to differentiate the iron oxide phases present within the nanoparticles, which could be possible through the electron energy loss spectroscopy method [92].

Although it might result in high agglomeration degrees, high surface areas of iron oxide nanoparticles are essential for ensuring the desired application [83, 87]. For example, in waste or pollutant removal applications, iron oxide nanoparticles must possess high surface areas to increase capture and immobilization efficiency [93]. Surface area is determined through straightforward gas sorption techniques, such as the BET analysis [84].

The surface charge can be correlated with the colloidal stability and interactions of the nanoparticles. Specifically, the interactions of iron oxide nanoparticles within the biological fluids will determine the formation of the protein corona on their surface and, thus, the probability of cellular uptake [84]. Generally, the surface charge of nanoparticles is measured through a zeta potentiometer by applying a voltage to the samples. The results are given in terms of zeta potential ζ, which refers to the difference in the electric potential between the particle surrounding stationary charge layers and the potential of the solution [94–97]. Zeta potential values higher than +15 mV and lower than −15 mV are usually attributed to colloidally stable suspensions, as they generate electrostatic repulsions that are strong enough to counteract aggregation of the nanoparticles [82, 84, 85]. However, there are several parameters that could influence the zeta potential, such as the pH and ionic strength of the solvent or the presence of charged/uncharged molecules that can be adsorbed onto the surface of the nanoparticles [84].

Considering the extensive studies performed toward hyperthermia applications for controlled drug delivery and cancer therapy, an essential characteristic of iron oxide nanoparticles is their magnetic behavior. The magnetic properties of iron oxide nanoparticles directly depend upon the synthesis route and the size and shape of the obtained nanostructures [35]. Similar to other properties, the magnetic behavior of nanostructured materials is significantly different than those of the bulk materials, since the size decrease leads to changes from the multidomain to the single domain and, finally, to the superparamagnetic state [83]. Specifically, nanoparticles with superparamagnetic properties are characterized by negligible remanent magnetization and coercive field. Thus, when the external magnetic field is removed, the nanoparticles exhibit no magnetism. By contrast, ferro- and ferrimagnetic nanoparticles feature a magnetic hysteresis, associated with a remanent magnetization, thus requiring a coercive field for reverting the magnetization to zero [35, 98]. Furthermore, the magnetic behavior can also be influenced by the agglomeration and aggregation processes, which further favor dipole-dipole or exchange interactions [99]. There are several methods that can be utilized for the analysis of iron oxide nanoparticle magnetism, each associated with specific sensitivities [35].

4. Conclusions

Iron oxide nanoparticles have been intensively studied for a variety of applications within numerous fields, ranging from medicine and pharmaceutics to microelectronics and analytical chemistry. Since their utilization is continuously rising, the need for improving the currently available synthesis methods is fundamental. In this context, novel preparation routes must be explored to develop uniform and standardized iron oxide nanoparticles. Among the recently implemented strategies, microwave-assisted, microfluidics, and green synthesis methods have demonstrated an undoubted potential toward reaching this goal. Specifically, nanoparticles obtained through these methods were characterized by superior properties as compared to the co-precipitation counterparts. Furthermore, the advancements within the characterization techniques could further lead to new insights and potential improvements within this area. In this context, the most important techniques often include size, shape, structure, crystallinity, and magnetic behavior determinations. Therefore, the intensive research work investigating the synthesis of iron oxide nanoparticles must further continue.

Acknowledgements

We acknowledge the support of the research grant from the Romanian National Authority for Scientific Research and Innovation, UEFISCDI, project number TE 103, code: PN-III-P1-1.1-TE-2019-1450, entitled multifunctional lab-on-a-chip microfluidic platform for the fabrication of nanoparticles.

Conflict of interest

The authors declare no conflict of interest.

Notes/thanks/other declarations

We thank the esteemed group of researchers that have created the National Research Center for Micro and Nanomaterials from the University Politehnica of Bucharest and have considerably contributed to the development of numerous research studies.

Author details

Cristina Chircov[1,2] and Bogdan Stefan Vasile[1,2]*

1 Department of Science and Engineering of Oxide Materials and Nanomaterials, University Politehnica of Bucharest, Bucharest, Romania

2 National Research Center for Micro and Nanomaterials, University Politehnica of Bucharest, Bucharest, Romania

*Address all correspondence to: bogdan.vasile@upb.ro

References

[1] Gahlot P, Aboudi K, Ahmed B, Tawfik A, Khan AA, Khursheed A, et al. Chapter 9—Direct interspecies electron transfer (DIET) via conductive materials in anaerobic digestion of organic wastes. In: Tyagi V, Aboudi K, editors. Clean Energy and Resources Recovery. Amsterdam, The Netherlands: Elsevier; 2021. pp. 227-252

[2] Ahmadi M. 16—Iron oxide nanoparticles for delivery purposes. In: Mozafari M, editor. Nanoengineered Biomaterials for Advanced Drug Delivery. Amsterdam, The Netherlands: Elsevier; 2020. p. 373-393

[3] Mishra M, Chun D-M. α-Fe_2O_3 as a photocatalytic material: A review. Applied Catalysis A: General. 2015;**498**:126-141. DOI: 10.1016/j.apcata.2015.03.023

[4] Schwertmann U, Taylor RM. Iron oxides. Minerals in soil environments. 1989;**1**:379-438

[5] Claudio C, Iorio E, Liu Q, Jiang Z, Barrón V. Iron oxide nanoparticles in soils: Environmental and agronomic importance. Journal of Nanoscience and Nanotechnology. 2017;**17**(7):4449-4460

[6] Von der Heyden B, Roychoudhury A, Myneni S. Iron-rich nanoparticles in natural aquatic environments. Minerals. 2019;**9**(5):287

[7] Ezealigo US, Ezealigo BN, Aisida SO, Ezema FI. Iron oxide nanoparticles in biological systems: Antibacterial and toxicology perspective. JCIS Open. 2021;**4**:100027. DOI: 10.1016/j.jciso.2021.100027

[8] Moura RP, Sousa F, Almeida A, Pinto S, Sarmento B. 13—Theranostic biomaterials for regulation of the blood–brain barrier. In: Cui W, Zhao X, editors. Theranostic Bionanomaterials. Amsterdam, The Netherlands: Elsevier; 2019. p. 303-319

[9] de Jesus PdC C, Pellosi DS, Tedesco AC. Chapter 12—Magnetic nanoparticles: Applications in biomedical processes as synergic drug-delivery systems. In: Holban A-M, Grumezescu AM, editors. Materials for Biomedical Engineering. Amsterdam, The Netherlands: Elsevier; 2019. pp. 371-396

[10] Baumgartner J, Faivre D. Iron solubility, colloids and their impact on iron (oxyhydr)oxide formation from solution. Earth-Science Reviews. 2015;**150**:520-530

[11] Hiemstra T. Formation, stability, and solubility of metal oxide nanoparticles: Surface entropy, enthalpy, and free energy of ferrihydrite. Geochimica et Cosmochimica Acta. 2015;**158**:179-198

[12] Weatherill JS, Morris K, Bots P, Stawski TM, Janssen A, Abrahamsen L, et al. Ferrihydrite formation: The role of Fe_{13} Keggin clusters. Environmental Science and Technology. 2016;**50**(17):9333-9342

[13] Chernyshova IV, Ponnurangam S, Somasundaran P. Linking interfacial chemistry of CO_2 to surface structures of hydrated metal oxide nanoparticles: Hematite. Physical Chemistry Chemical Physics. 2013;**15**(18):6953-6964

[14] Singla A, Marwaha A, Marwaha S, Chavali M. Chapter 2—Green nanomaterials: Pioneering approach toward hyperthermia

treatment. In: Kharisov B, Kharissova O, editors. Handbook of Greener Synthesis of Nanomaterials and Compounds. Amsterdam, The Netherlands: Elsevier; 2021. pp. 487-502

[15] Jean-Pierre J, Elisabeth T, Corinne C. Iron oxides: From molecular clusters to solid. A nice example of chemical versatility. Comptes Rendus Geoscience. 2006;**338**:488-497

[16] Miyazaki T. 4—Organic modification of magnetite nanoparticles for biomedical applications. In: Osaka A, Narayan R, editors. Bioceramics. Amsterdam, The Netherlands: Elsevier; 2021. pp. 77-82

[17] Daoush WM. Co-precipitation and magnetic properties of magnetite nanoparticles for potential biomedical applications. Journal of Nanomedicine Research. 2017;**5**(3):00118

[18] Ganapathe LS, Mohamed MA, Mohamad Yunus R, Berhanuddin DD. Magnetite (Fe3O4) nanoparticles in biomedical application: From synthesis to surface functionalisation. Magnetochemistry. 2020;**6**(4):68. DOI: 10.3390/magnetochemistry6040068

[19] Ortega G, Reguera E. Chapter 13—Biomedical applications of magnetite nanoparticles. In: Holban A-M, Grumezescu AM, editors. Materials for Biomedical Engineering. Amsterdam, The Netherlands: Elsevier; 2019. pp. 397-434

[20] Sajid M, Płotka-Wasylka J. Nanoparticles: Synthesis, characteristics, and applications in analytical and other sciences. Microchemical Journal. 2020;**154**:104623. DOI: 10.1016/j.microc.2020.104623

[21] Goswami L, Kim K-H, Deep A, Das P, Bhattacharya SS, Kumar S, et al. Engineered nano particles: Nature, behavior, and effect on the environment. Journal of Environmental Management. 2017;**196**:297-315

[22] Tanaka S, Kaneti YV, Septiani NLW, Dou SX, Bando Y, Hossain MSA, et al. A review on iron oxide-based nanoarchitectures for biomedical, energy storage, and environmental applications. Small Methods. 2019;**3**(5):1800512

[23] Gao L, Zhuang J, Nie L, Zhang J, Zhang Y, Gu N, et al. Intrinsic peroxidase-like activity of ferromagnetic nanoparticles. Nature Nanotechnology. 2007;**2**(9):577-583. DOI: 10.1038/nnano.2007.260

[24] Huang X-L, Zhang J-Z. Hydrolysis of glucose-6-phosphate in aged, acid-forced hydrolysed nanomolar inorganic iron solutions—An inorganic biocatalyst? RSC Advances. 2012;**2**(1):199-208

[25] Huang X-L. Hydrolysis of phosphate esters catalyzed by inorganic iron oxide nanoparticles acting as biocatalysts. Astrobiology. 2018;**18**(3):294-310

[26] Wei H, Wang E. Fe_3O_4 magnetic nanoparticles as peroxidase mimetics and their applications in H_2O_2 and glucose detection. Analytical Chemistry. 2008;**80**(6):2250-2254

[27] Lin Y, Ren J, Qu X. Catalytically active nanomaterials: A promising candidate for artificial enzymes. Accounts of Chemical Research. 2014;**47**(4):1097-1105

[28] Zhou Y, Liu B, Yang R, Liu J. Filling in the gaps between nanozymes and enzymes: Challenges and opportunities. Bioconjugate Chemistry. 2017;**28**(12):2903-2909

[29] Huang X-L. Iron oxide nanoparticles: An inorganic phosphatase. In:

Nanocatalysts. London, United Kingdom: IntechOpen; 2019

[30] Akbarzadeh A, Samiei M, Davaran S. Magnetic nanoparticles: Preparation, physical properties, and applications in biomedicine. Nanoscale Research Letters. 2012;**7**(1):144. DOI: 10.1186/1556-276X-7-144

[31] Gatoo M, Naseem S, Arfat M, Dar A, Qasim K, Zubair S. Physicochemical properties of nanomaterials: Implication in associated toxic manifestations. BioMed Research International. 2014;**2014**:498420. DOI: 10.1155/2014/498420

[32] Hui BH, Salimi MN. Production of iron oxide nanoparticles by co-precipitation method with optimization studies of processing temperature, pH and stirring rate. IOP Conference Series: Materials Science and Engineering. 2020;**743**:012036. DOI: 10.1088/1757-899x/743/1/012036

[33] Zhang J, Tang H, Liu Z, Chen B. Effects of major parameters of nanoparticles on their physical and chemical properties and recent application of nanodrug delivery system in targeted chemotherapy. International Journal of Nanomedicine. 2017;**12**:8483

[34] LaGrow AP, Besenhard MO, Hodzic A, Sergides A, Bogart LK, Gavriilidis A, et al. Unravelling the growth mechanism of the co-precipitation of iron oxide nanoparticles with the aid of synchrotron X-Ray diffraction in solution. Nanoscale. 2019;**11**(14):6620-6628. DOI: 10.1039/C9NR00531E

[35] Ali A, Shah T, Ullah R, Zhou P, Guo M, Ovais M, et al. Review on recent progress in magnetic nanoparticles: Synthesis, characterization, and diverse applications. Frontiers in Chemistry. 2021;**9**(548):629054. DOI: 10.3389/fchem.2021.629054

[36] Schwaminger SP, Syhr C, Berensmeier S. Controlled synthesis of magnetic iron oxide nanoparticles: Magnetite or maghemite? Crystals. 2020;**10**(3):214. DOI: 10.3390/cryst10030214

[37] Guo H, Barnard AS. Naturally occurring iron oxide nanoparticles: Morphology, surface chemistry and environmental stability. Journal of Materials Chemistry A. 2013;**1**(1):27-42

[38] Niculescu A-G, Chircov C, Grumezescu AM. Magnetite nanoparticles: Synthesis methods—A comparative review. Methods. 2021. (In Press). DOI: 10.1016/j.ymeth.2021.04.018

[39] Singh JP, Kumar M, Sharma A, Pandey G, Chae KH, Lee S. Bottom-up and top-down approaches for MgO. In: Sonochemical Reactions. London, United Kingdom: IntechOpen; 2020

[40] Priyadarshana G, Kottegoda N, Senaratne A, de Alwis A, Karunaratne V. Synthesis of magnetite nanoparticles by top-down approach from a high purity ore. Journal of Nanomaterials. 2015;**2015**:317312. DOI: 10.1155/2015/317312

[41] Aryal S, Park H, Leary JF, Key J. Top-down fabrication-based nano/microparticles for molecular imaging and drug delivery. International Journal of Nanomedicine. 2019;**14**:6631

[42] Baig N, Kammakakam I, Falath W. Nanomaterials: A review of synthesis methods, properties, recent progress, and challenges. Materials Advances. 2021;**2**(6):1821-1871

[43] Mihai AD, Chircov C, Grumezescu AM, Holban AM. Magnetite nanoparticles and essential oils systems for advanced antibacterial therapies. International Journal of Molecular

Sciences. 2020;**21**(19):7355. DOI: 10.3390/ijms21197355

[44] Rashid H, Mansoor MA, Haider B, Nasir R, Abd Hamid SB, Abdulrahman A. Synthesis and characterization of magnetite nano particles with high selectivity using in-situ precipitation method. Separation Science and Technology. 2020;**55**(6):1207-1215. DOI: 10.1080/01496395.2019.1585876

[45] Ansari SAMK, Ficiarà E, Ruffinatti FA, Stura I, Argenziano M, Abollino O, et al. Magnetic iron oxide nanoparticles: Synthesis, characterization and functionalization for biomedical applications in the central nervous system. Materials. 2019;**12**(3):465

[46] Wallyn J, Anton N, Vandamme TF. Synthesis, principles, and properties of magnetite nanoparticles for in vivo imaging applications-a review. Pharmaceutics. 2019;**11**(11):601. DOI: 10.3390/pharmaceutics11110601

[47] Liu S, Yu B, Wang S, Shen Y, Cong H. Preparation, surface functionalization and application of Fe3O4 magnetic nanoparticles. Advances in Colloid and Interface Science. 2020;**281**:102165. DOI: 10.1016/j.cis.2020.102165

[48] Wu W, Jiang CZ, Roy VA. Designed synthesis and surface engineering strategies of magnetic iron oxide nanoparticles for biomedical applications. Nanoscale. 2016;**8**(47): 19421-19474. DOI: 10.1039/c6nr07542h

[49] Dahiya MS, Tomer VK, Duhan S. 31—Metal–ferrite nanocomposites for targeted drug delivery. In: Inamuddin, Asiri AM, Mohammad A, editors. Applications of Nanocomposite Materials in Drug Delivery. Cambridge, United Kingdom: Woodhead Publishing; 2018. pp. 737-760

[50] Morsali A, Hashemi L. Chapter 2—Nanoscale coordination polymers: Preparation, function and application. In: Ruiz-Molina D, van Eldik R, editors. Advances in Inorganic Chemistry. 76: London, United Kingdom: Academic Press; 2020. p. 33-72.

[51] Gupta D, Jamwal D, Rana D, Katoch A. 26—Microwave synthesized nanocomposites for enhancing oral bioavailability of drugs. In: Inamuddin, Asiri AM, Mohammad A, editors. Applications of Nanocomposite Materials in Drug Delivery. Cambridge, United Kingdom: Woodhead Publishing; 2018. pp. 619-632

[52] Bandyopadhyay D, Banik BK. Chapter 3—Microwave-assisted synthesis of medicinally privileged heterocycles. In: Brahmachari G, editor. Green Synthetic Approaches for Biologically Relevant Heterocycles. 2nd ed. Amsterdam, The Netherlands: Elsevier; 2021. pp. 49-110

[53] Aivazoglou E, Metaxa E, Hristoforou E. Microwave-assisted synthesis of iron oxide nanoparticles in biocompatible organic environment. AIP Advances. 2018;**8**(4):048201. DOI: 10.1063/1.4994057

[54] Kostyukhin EM, Kustov LM. Microwave-assisted synthesis of magnetite nanoparticles possessing superior magnetic properties. Mendeleev Communications. 2018;**28**(5):559-561. DOI: 10.1016/j.mencom.2018.09.038

[55] Chircov C, Matei M-F, Neacşu IA, Vasile BS, Oprea O-C, Croitoru A-M, et al. Iron oxide–silica core–shell nanoparticles functionalized with essential oils for antimicrobial therapies. Antibiotics. 2021;**10**(9):1138

[56] Brollo MEF, Veintemillas-Verdaguer S, Salván CM, MdP M. Key parameters on the microwave assisted synthesis of magnetic nanoparticles for MRI contrast agents. Contrast Media & Molecular Imaging. 2017;**2017**:8902424. DOI: 10.1155/2017/8902424

[57] Chircov C, Bîrcă AC, Grumezescu AM, Vasile BS, Oprea O, Nicoară AI, et al. Synthesis of magnetite nanoparticles through a lab-on-chip device. Materials. 2021;**14**(19):5906. DOI: 10.3390/ma14195906

[58] Chircov C, Bîrcă AC, Grumezescu AM, Andronescu E. Biosensors-on-chip: An up-to-date review. Molecules. 2020;**25**(24):6013. DOI: 10.3390/molecules25246013

[59] Niculescu A-G, Chircov C, Bîrcă AC, Grumezescu AM. Nanomaterials synthesis through microfluidic methods: An updated overview. Nanomaterials. 2021;**11**(4):864. DOI: 10.3390/nano11040864

[60] Abedini-Nassab R, Pouryosef Miandoab M, Şaşmaz M. Microfluidic synthesis, control, and sensing of magnetic nanoparticles: A review. Micromachines (Basel). 2021;**12**(7):768. DOI: 10.3390/mi12070768

[61] Thu VT, Mai AN, Le The T, Van Trung H, Thu PT, Tien BQ, et al. Fabrication of PDMS-based microfluidic devices: Application for synthesis of magnetic nanoparticles. Journal of Electronic Materials. 2016;**45**(5):2576-2581. DOI: 10.1007/s11664-016-4424-6

[62] Bemetz J, Wegemann A, Saatchi K, Haase A, Häfeli UO, Niessner R, et al. Microfluidic-based synthesis of magnetic nanoparticles coupled with miniaturized NMR for online relaxation studies. Analytical Chemistry. 2018;**90**(16):9975-9982. DOI: 10.1021/acs.analchem.8b02374

[63] Aşık MD, Kaplan M, Çetin B, Sağlam N. Synthesis of iron oxide core chitosan nanoparticles in a 3D printed microfluidic device. Journal of Nanoparticle Research. 2021;**23**(3):62. DOI: 10.1007/s11051-021-05171-y

[64] Siavashy S, Soltani M, Ghorbani-Bidkorbeh F, Fallah N, Farnam G, Mortazavi SA, et al. Microfluidic platform for synthesis and optimization of chitosan-coated magnetic nanoparticles in cisplatin delivery. Carbohydrate Polymers. 2021;**265**:118027. DOI: 10.1016/j.carbpol.2021.118027

[65] Zhang D, Ma X-l, Gu Y, Huang H, Zhang G-w. Green synthesis of metallic nanoparticles and their potential applications to treat cancer. Frontiers in Chemistry. 2020;**8**:799. DOI: 10.3389/fchem.2020.00799

[66] Ahmad S, Munir S, Zeb N, Ullah A, Khan B, Ali J, et al. Green nanotechnology: A review on green synthesis of silver nanoparticles—An ecofriendly approach. International Journal of Nanomedicine. 2019;**14**:5087

[67] Gour A, Jain NK. Advances in green synthesis of nanoparticles. Artificial Cells, Nanomedicine, and Biotechnology. 2019;**47**(1):844-851. DOI: 10.1080/21691401.2019.1577878

[68] Dikshit PK, Kumar J, Das AK, Sadhu S, Sharma S, Singh S, et al. Green synthesis of metallic nanoparticles: Applications and limitations. Catalysts. 2021;**11**(8):902

[69] Naikoo GA, Mustaqeem M, Hassan IU, Awan T, Arshad F, Salim H, et al. Bioinspired and green synthesis of

nanoparticles from plant extracts with antiviral and antimicrobial properties: A critical review. Journal of Saudi Chemical Society. 2021;**25**(9):101304. DOI: 10.1016/j.jscs.2021.101304

[70] Jadoun S, Arif R, Jangid NK, Meena RK. Green synthesis of nanoparticles using plant extracts: A review. Environmental Chemistry Letters. 2021;**19**(1):355-374. DOI: 10.1007/s10311-020-01074-x

[71] Shafey AME. Green synthesis of metal and metal oxide nanoparticles from plant leaf extracts and their applications: A review. Green Processing and Synthesis. 2020;**9**(1):304-339. DOI: 10.1515/gps-2020-0031

[72] Yadwade R, Kirtiwar S, Ankamwar B. A review on green synthesis and applications of iron oxide nanoparticles. Journal of Nanoscience and Nanotechnology. 2021;**21**(12):5812-5834

[73] Priya N, Kaur K, Sidhu AK. Green synthesis: An eco-friendly route for the synthesis of iron oxide nanoparticles. Frontiers in Nanotechnology. 2021;**3**(47):655062. DOI: 10.3389/fnano.2021.655062

[74] Lakshminarayanan S, Shereen MF, Niraimathi KL, Brindha P, Arumugam A. One-pot green synthesis of iron oxide nanoparticles from Bauhinia tomentosa: Characterization and application towards synthesis of 1,3 diolein. Scientific Reports. 2021;**11**(1):8643. DOI: 10.1038/s41598-021-87960-y

[75] Bibi I, Nazar N, Ata S, Sultan M, Ali A, Abbas A, et al. Green synthesis of iron oxide nanoparticles using pomegranate seeds extract and photocatalytic activity evaluation for the degradation of textile dye. Journal of Materials Research and Technology. 2019;**8**(6):6115-6124. DOI: 10.1016/j.jmrt.2019.10.006

[76] Razack SA, Suresh A, Sriram S, Ramakrishnan G, Sadanandham S, Veerasamy M, et al. Green synthesis of iron oxide nanoparticles using Hibiscus rosa-sinensis for fortifying wheat biscuits. SN Applied Sciences. 2020;**2**(5):898. DOI: 10.1007/s42452-020-2477-x

[77] Niraimathee VA, Subha V, Ramaswami Sachidanandan ER, Renganathan S. Green synthesis of iron oxide nanoparticles from Mimosa pudica root extract. International Journal of Environment and Sustainable Development. 2016;**15**:227. DOI: 10.1504/IJESD.2016.077370

[78] Paul S, Aka T, Ashaduzzaman M, Rahaman MM, Otun S. Green synthesis of iron oxide nanoparticle using Carica papaya leaf extract: Application for photocatalytic degradation of remazol yellow RR dye and antibacterial activity. Heliyon. 2020;**6**:e04603. DOI: 10.1016/j.heliyon.2020.e04603

[79] Patiño-Ruiz DA, Meramo-Hurtado SI, González-Delgado ÁD, Herrera A. Environmental sustainability evaluation of iron oxide nanoparticles synthesized via green synthesis and the coprecipitation method: A comparative life cycle assessment study. ACS Omega. 2021;**6**(19):12410-12423. DOI: 10.1021/acsomega.0c05246

[80] Üstün E, Önbaş SC, Çelik SK, Ayvaz MÇ, Şahin N. Green synthesis of iron oxide nanoparticles by using Ficus carica leaf extract and its antioxidant activity. Biointerface Research in Applied Chemistry. 2022;**2021**(12):2108-2116

[81] Devi HS, Boda MA, Shah MA, Parveen S, Wani AH. Green synthesis of iron oxide nanoparticles using

Platanus orientalis leaf extract for antifungal activity. Green Processing and Synthesis. 2019;**8**(1):38-45. DOI: 10.1515/gps-2017-0145

[82] Anu Mary Ealia S, Saravanakumar MP. A review on the classification, characterisation, synthesis of nanoparticles and their application. IOP Conference Series: Materials Science and Engineering. 2017;**263**:032019. DOI: 10.1088/1757-899x/263/3/032019

[83] Mourdikoudis S, Pallares RM, Thanh NTK. Characterization techniques for nanoparticles: Comparison and complementarity upon studying nanoparticle properties. Nanoscale. 2018;**10**(27):12871-12934. DOI: 10.1039/C8NR02278J

[84] Modena MM, Rühle B, Burg TP, Wuttke S. Nanoparticle characterization: What to measure? Advanced Materials. 2019;**31**(32):1901556. DOI: 10.1002/adma.201901556

[85] Bélteky P, Rónavári A, Zakupszky D, Boka E, Igaz N, Szerencsés B, et al. Are smaller nanoparticles always better? understanding the biological effect of size-dependent silver nanoparticle aggregation under biorelevant conditions. International Journal of Nanomedicine. 2021;**16**:3021

[86] Arno MC, Inam M, Weems AC, Li Z, Binch ALA, Platt CI, et al. Exploiting the role of nanoparticle shape in enhancing hydrogel adhesive and mechanical properties. Nature Communications. 2020;**11**(1):1420. DOI: 10.1038/s41467-020-15206-y

[87] Gomathi T, Rajeshwari K, Kanchana V, Sudha PN, Parthasarathy K. Impact of nanoparticle shape, size, and properties of the sustainable nanocomposites. In: Inamuddin, Thomas S, Kumar Mishra R, Asiri AM, editors. Sustainable Polymer Composites and Nanocomposites. Cham: Springer International Publishing; 2019. pp. 313-336

[88] Boselli L, Lopez H, Zhang W, Cai Q, Giannone VA, Li J, et al. Classification and biological identity of complex nano shapes. Communications Materials. 2020;**1**(1):35. DOI: 10.1038/s43246-020-0033-2

[89] Khan I, Saeed K, Khan I. Nanoparticles: Properties, applications and toxicities. Arabian Journal of Chemistry. 2019;**12**(7):908-931. DOI: 10.1016/j.arabjc.2017.05.011

[90] Vorontsov AV, Tsybulya SV. Influence of nanoparticles size on XRD patterns for small monodisperse nanoparticles of Cu0 and TiO2 anatase. Industrial and Engineering Chemistry Research. 2018;**57**(7):2526-2536. DOI: 10.1021/acs.iecr.7b04480

[91] Mukundan V, Shan S, Zhong C-J, Malis O. Effect of chemical composition on the nanoscale ordering transformations of physical mixtures of Pd and Cu nanoparticles. Journal of Nanomaterials. 2018;**2018**:9087320. DOI: 10.1155/2018/9087320

[92] Cavé L, Al T, Loomer D, Cogswell S, Weaver L. A STEM/EELS method for mapping iron valence ratios in oxide minerals. Micron. 2006;**37**(4):301-309. DOI: 10.1016/j.micron.2005.10.006

[93] Koopmans GF, Hiemstra T, Vaseur C, Chardon WJ, Voegelin A, Groenenberg JE. Use of iron oxide nanoparticles for immobilizing phosphorus in-situ: Increase in soil reactive surface area and effect on soluble phosphorus. Science of the Total Environment. 2020;**711**:135220. DOI: 10.1016/j.scitotenv.2019.135220

[94] Ismail AF, Khulbe KC, Matsuura T. Chapter 3—RO membrane characterization. In: Ismail AF, Khulbe KC, Matsuura T, editors. Reverse Osmosis. Amsterdam, The Netherlands: Elsevier; 2019. pp. 57-90

[95] Kumar CV, Pattammattel A. Chapter 8—Graphene composites with proteins and biologics. In: Kumar CV, Pattammattel A, editors. Introduction to Graphene. Amsterdam, The Netherlands: Elsevier; 2017. pp. 155-186

[96] Yamashita Y, Sakamoto K. Chapter 38—Structural analysis of formulations. In: Sakamoto K, Lochhead RY, Maibach HI, Yamashita Y, editors. Cosmetic Science and Technology. Amsterdam: Elsevier; 2017. pp. 635-655

[97] Nasrollahzadeh M, Shafiei N, Soleimani F, Nezafat Z, Soheili Bidgoli NS. Chapter 10—Physicochemical characterization of biopolymer-based metal nanoparticles. In: Nasrollahzadeh M, editor. Biopolymer-Based Metal Nanoparticle Chemistry for Sustainable Applications. Amsterdam, The Netherlands: Elsevier; 2021. pp. 317-478

[98] Maldonado-Camargo L, Unni M, Rinaldi C. Magnetic characterization of iron oxide nanoparticles for biomedical applications. Methods in Molecular Biology. 2017;**1570**:47-71. DOI: 10.1007/978-1-4939-6840-4_4

[99] Gutiérrez L, de la Cueva L, Moros M, Mazarío E, de Bernardo S, de la Fuente JM, et al. Aggregation effects on the magnetic properties of iron oxide colloids. Nanotechnology. 2019;**30**(11):112001. DOI: 10.1088/1361-6528/aafbff

Section 3

Characterization

Chapter 3

Enzyme-Like Property (Nanozyme) of Iron Oxide Nanoparticles

Lizeng Gao

Abstract

Iron oxide nanoparticles perform biological activity under physiological conditions. They exhibit enzyme-like properties that catalyze redox reactions mediated by natural enzymes of oxidoreductase and are classified into a typical of nanozymes that are defined as nanomaterials with enzyme-like activities. In addition, iron oxide nanoparticles widely exist in biological system, such as magnetosome and ferritin that not only regulate iron metabolism, but also regulate ROS homostasis. The enzyme-like properties of iron oxide nanoparticles render them with broad biomedical applications including immunoassay, biosensor, antimicrobial, anti-tumor, antioxidant. Taken together, iron oxide nanoparticles are bioactive materials and may perform particular biological function in life activity.

Keywords: iron oxide, enzyme-like property, nanozyme, ROS regulation, biological function

1. Introduction

Iron oxide nanoparticles are of typical nanomaterials which can be synthesized using chemical methods or by made from iron oxide minerals. Iron oxide nanoparticles usually include several phases such as magnetite (Fe_3O_4), maghemite (γ-Fe_2O_3), and ferrihydrite (Fh) [1–3]. The surface structure, crystal phase and facet, shape as well as size dependence make them have various functions [4, 5].

Due to excellent magnetic property, iron oxide nanoparticles have been broadly used in biomedical field, such as magnetic separation of biosamples (nucleic acids, proteins, cells), drug delivery, tumor hyperthermia. Among these applications, iron oxide nanoparticles are assumed as biological inert, even if they have been used as Fenton catalysts for advanced oxidation in waste treatment. In recent years, enzyme-like properties of iron oxide nanoparticles have drawn more attention. In 2007, Yan group found that these nanoparticles performed intrinsic peroxidase (POD)-like activity [6]. Since then, many nanoparticles are found with enzyme-like properties, which boost the development of nanozymes [7, 8]. In particular, iron oxide nanoparticles are found with multiple enzyme-like activities. For instance, Xiaolan Huang found that iron oxide nanoparticles exhibit phosphatase activity that can hydrolyze phosphate ester. Currently, iron oxide nanoparticles are classified as one typical nanozymes and extend their biomedical applications such as antibacterial, antiviral, antitumor, antioxidant, immune regulation.

In addition, iron oxide nanoparticles are also found in biological system, such as magnetosomes of bacteria and ferritin, which are classified as natural nanozymes. These biological iron oxide nanoparticles may be involved in metabolic processes and contribute to life evolution. Thus, the studies on enzyme-like property of iron oxide nanoparticles not only have important significance in extending their biomedical applications, but also provide clues for origin of life. In this chapter, we will summarize the cutting-edged progress in the field of iron oxide nanoparticles related to biological properties (mainly for enzyme-like catalysis) and highlight the significance of such properties in biomedicine and nanobiology.

2. Enzyme-like activities of iron oxide nanoparticles

Iron oxide nanoparticles perform enzyme-like activities that can catalyze the biochemical reactions mediated by natural enzymes under mild condition. The fundamental rationale is that iron serves as critical cofactor in the active center of many natural enzymes. Currently, more than 80 types of natural enzymes are found with iron as cofactors in the form of hemin, coordinated single iron or di-iron, iron-sulfur clusters, which perform the activities ranging from oxidoreductases to nitrogenase. Since iron oxide nanoparticles are rich in ferrous and ferric iron, it is rational to speculate that these nanoparticles mimic the activities of iron-containing enzymes.

It should be noted that the structure of iron oxide nanoparticles is quite different to that of natural enzymes. It is rigid crystal structure in iron oxide nanoparticles, and the active sites may locate on the surface of the nanoparticle. Such inorganic structure endows iron oxide nanoparticles with superior stability and high activity (multiple active sites in single nanoparticle), which makes them suitable for applications under unfriendly environments for natural enzymes. Below, we will introduce the enzyme-like activities reported in the recent decade, in particular in the view of nanozymes.

2.1 Peroxidase-like activity

In 2007, Yan group for the first time reported that ferromagnetic iron oxide (Fe_3O_4) nanoparticles perform intrinsic peroxidase-like activity that can catalyze colorimetric reaction in the presence of hydrogen peroxide (H_2O_2) under acidic condition (pH 3–6.5) (**Figure 1**) [6, 9]. In this reaction, H_2O_2 is converted to free radicals (•OH) as intermediate. The catalytic behaviors including substrates, optimal pH, and temperature of iron oxide nanoparticles are all similar as those of horseradish peroxidase (HRP). In particular, enzymatic kinetics assay showed that iron oxide nanoparticles fit Michaelis–Menten equation and follow ping-pong mechanism, which confirmed that iron oxide nanoparticles are mimic of HRP. For nanoparticles with the size at 300 nm in diameter, the catalytic efficiency of a single nanoparticle is comparable with a single HRP molecule. However, this does not mean that iron in the nanoparticle has higher activity than that in HRP. Each HRP molecule only has one active site and one iron. In contrast, there may be multiple active sites on the surface of one iron oxide nanoparticle, strongly related to crystal structures, exposed facets, defects, and chemical modifications (**Figure 2**).

Xingfa Gao group used density functional theory calculations to investigate the peroxidase-mimetic mechanisms for a series of iron oxide nanosurfaces [10]. They proposed that the activity of these iron oxide nanoparticles mimicked that of POD by

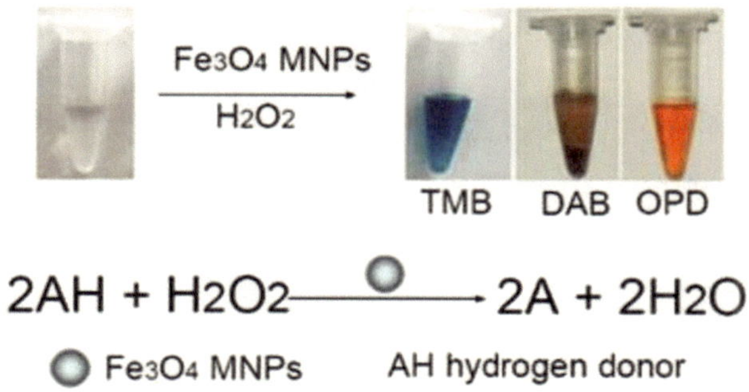

Figure 1.
Iron oxide nanoparticles with peroxidase-like activity catalyze colorimetric reaction mediated by HRP [6]. Copyright 2007, Springer Nature.

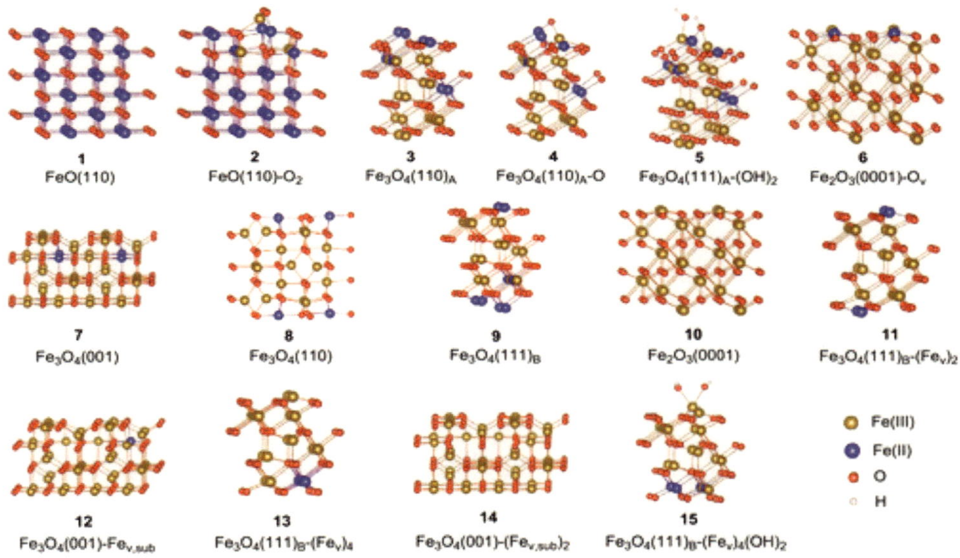

Figure 2.
Diagrams of iron oxide slabs with different crystal structures, exposed facets, defects, and chemical modifications [10]. Copyright 2020, ACS.

following a three-step mechanism in which chemisorption of H_2O_2 was absorbed onto the surface to form two hydroxyl adsorbates and two subsequent reduction processes to remove the hydroxyl groups from the surface. The POD-like catalyses of all iron oxide surfaces proceeded *via* almost the same mechanism. The properties of iron oxides tuned the energy barrier heights of reaction steps and thus determined which step to be the rate-determining step, resulting in different catalytic kinetics and activity for the surfaces (**Figure 3**). These theoretical analyses help to understand the relationship between structure and activity of iron oxide nanoparticles.

Compared with natural enzymes, iron oxide nanoparticles exhibit high stability to non-physiological conditions such as low or high temperature, acidic or basic pH, organic solvents. In addition, the peroxidase-like activity of iron oxide nanoparticles is tunable by adjusting their size, morphology, facets, defects, or surface modifications. Similar to natural enzymes, the activity of iron oxide nanoparticles

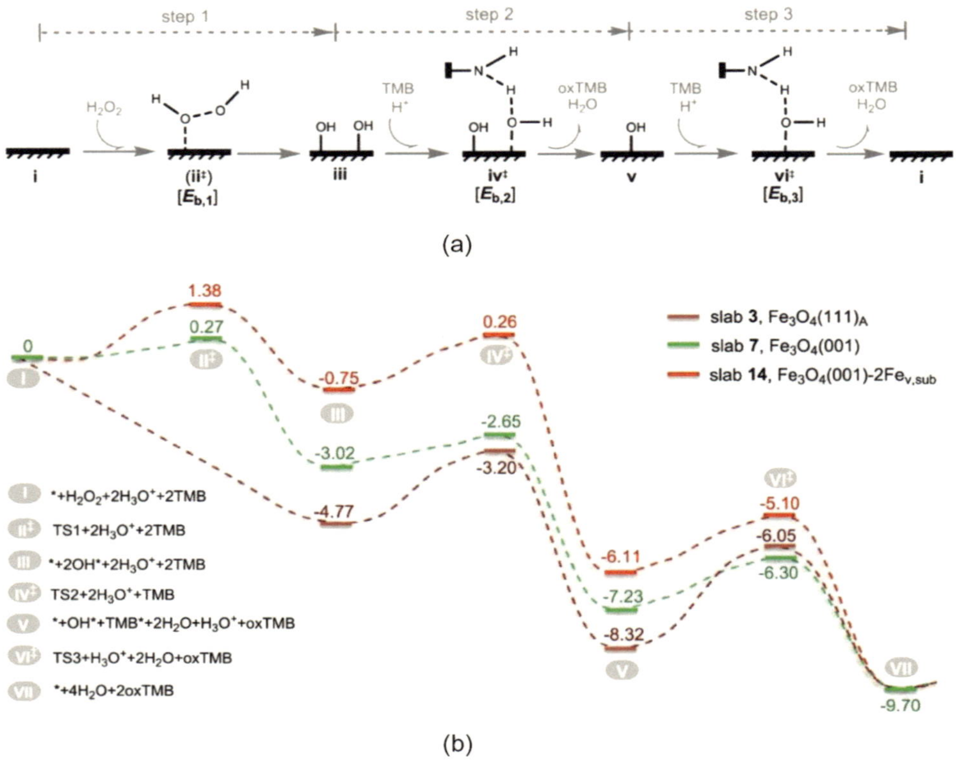

Figure 3.
Mechanism and kinetics of POD-like reactions catalyzed by iron oxides as determined from DFT calculations [10]. (a) Proposed mechanism of POD-mimetic catalysis of iron oxide slabs. (b) Relative energy values (in eV) for key intermediates and transition states involved in the catalytic cycles. Copyright 2020, ACS.

also can be affected by activators or inhibitors [11]. However, the specific peroxidase-like activity of pure iron oxide nanoparticles is in the range of several U/mg, which is much lower than that of HRP with specific activity >150 U/mg [12]. To enhance the specific activity, iron oxide nanoparticles can by hybridized with other nanovectors to form nanocomplexes. Juewen Liu group used molecular imprinting to modify the surface and achieved selective catalysis for the substrates of TMB and ABTS. They found that introducing charged monomers led to nearly 100-fold specificity for the imprinted substrate over the nonimprinted compared with that of bare Fe_3O_4 [13].

2.2 Catalase-like activity

In addition to peroxidase-like activity, iron oxide nanoparticles were found with catalase-like activity under neutral pH by Gu's group [14]. Using electron spin resonance spectroscopy, they found that both Fe_3O_4 and γ-Fe_2O_3 nanoparticles decomposed H_2O_2 into hydroxyl radicals under acidic condition (pH < 6.5), showing peroxidase-like activity (Fe_3O_4 > γ-Fe_2O_3). However, H_2O_2 was decomposed into H_2O and O_2 under neutral pH (pH 7.4) condition by the two nanoparticles, demonstrating catalase-like activity. These results indicated that the enzyme-like activities of iron oxide nanoparticles are pH-dependent; that is, peroxidase-like activity is dominant at acidic pH and catalase-like activity is dominant at neutral pH (**Figure 4**).

Mover, ferrihydrite, a precursor for most iron oxides, was found with catalase-like activity by Fan's group [15]. They found that among the 10 forms of iron oxide nanoparticles, 2-line ferrihydrite exhibited the highest catalase-like activity in the pH range of 4.0–8.7, but no peroxidase-like and superoxide dismutase-like activity. The structure-activity studies indicated that the surface iron-associated hydroxyl groups play a key role in catalase-like catalysis. Since natural catalase uses hemin as cofactor in active center, the catalase-like property of the previously mentioned iron oxide nanoparticles may be derived from the iron on the surface (**Figure 5**).

2.3 Superoxide dismutase-like activity

Inspired by some natural superoxide dismutase (SOD) using iron as cofactor, iron oxide nanoparticles are expected to perform SOD-like activity that converts superoxide ($O_2^{-\bullet}$) into O_2 and H_2O_2 or H_2O under basic pH (7 ~ 8). However, naked iron oxide nanoparticles exhibited quite low SOD-like activity. Gu et al. modified vitamin B2 on iron oxide nanoparticles and significantly improved the SOD-like activity, providing a reactive oxygen species (ROS)-scavenging ability [16] (**Figure 6**).

2.4 Oxidase-like activity

Inspired by some natural lipoxidase using iron as cofactor, iron oxide nanoparticles are expected to perform activity inducing lipid peroxidation. Tao Qin et al. incubated iron oxide nanoparticles with liposome at neutral and found that lipid peroxidation occurred by measuring MDA. This phenomenon was repeated using virus containing lipid envelope, which can disrupt viral integrity and degrade surface protein related to infecting host cells [17].

Of noted, although iron oxide nanoparticles exhibited four oxidoreductase-like activities [18], the catalytic efficiency of each activity is different, which may follow the order: peroxidase>catalase>SOD>lipoxidase. In addition, these activities show pH dependency. The pH range may have overlap, and thus, iron oxide nanoparticles may perform multiple activities simultaneously at a specific pH.

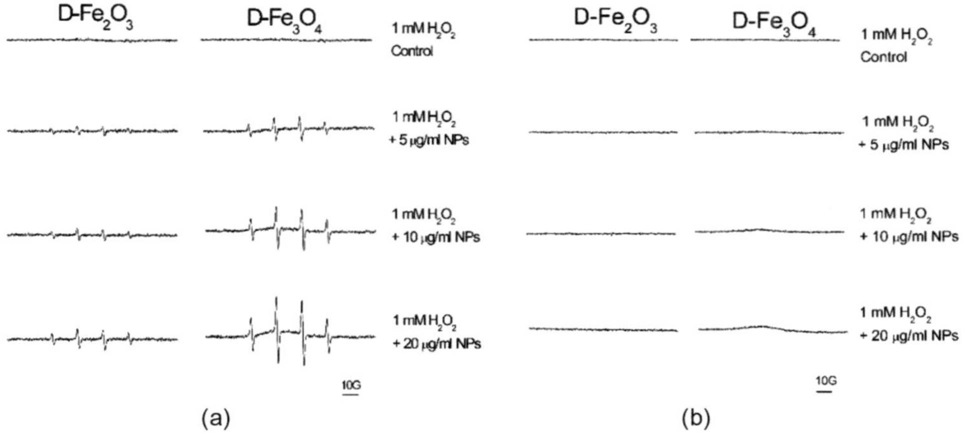

Figure 4.
ESR spectra subtraction of spin adduct DMPO/•OH [14]. All mixtures contained zero (control) or IONPs at different concentrations and 50 mM DMPO in (a) 100 mM acetate buffer (pH = 4.8) and (b) 50 mM PBS buffer (pH = 7.4). The reaction was initiated by adding 1 mM H_2O_2. Copyright 2012, ACS.

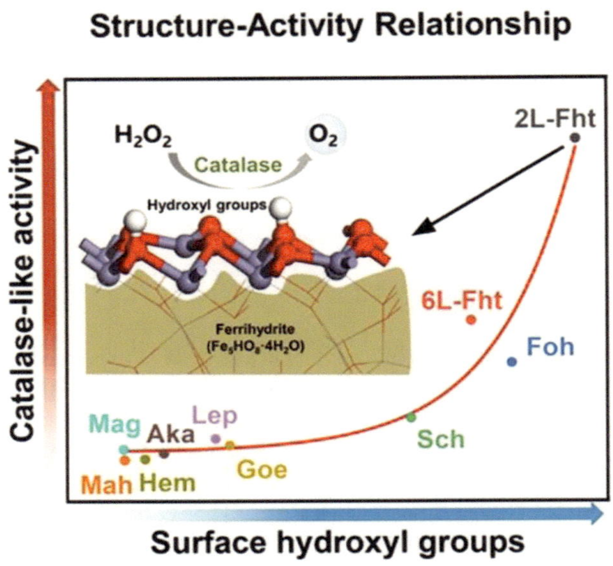

Figure 5.
The structure-activity relationship of ferrihydrite nanoparticles in catalase-like catalysis [15]. Copyright 2021, Elsevier.

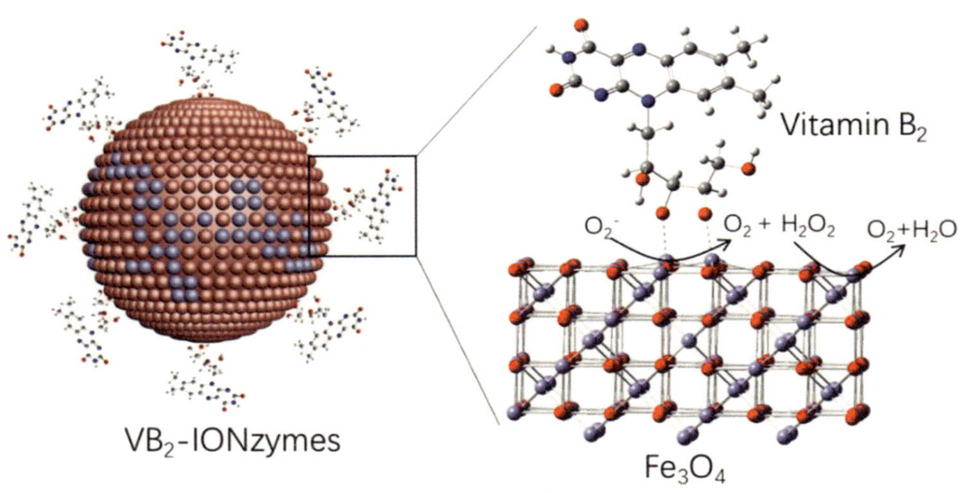

Figure 6.
Modification of VB2 to improve SOD-like activity of iron oxide nanoparticles [16]. Copyright 2020, Springer.

2.5 Phosphatase-like activity

Beside the previously mentioned oxidoreductase-like properties, Xiao-Lan Huang discovered that iron oxide nanoparticles exhibited the activity to catalyze the hydrolysis of phosphate ester with enzyme-like kinetics [19, 20]. The iron oxide nanoparticles prepared using a dialysis membrane tube (DMT) system led to the decrease of phosphate esters such as G6P, ATP, G2P and the increase of inorganic orthophosphate (Pi), indicating a catalytic effect on the hydrolysis reaction, which is mediated by

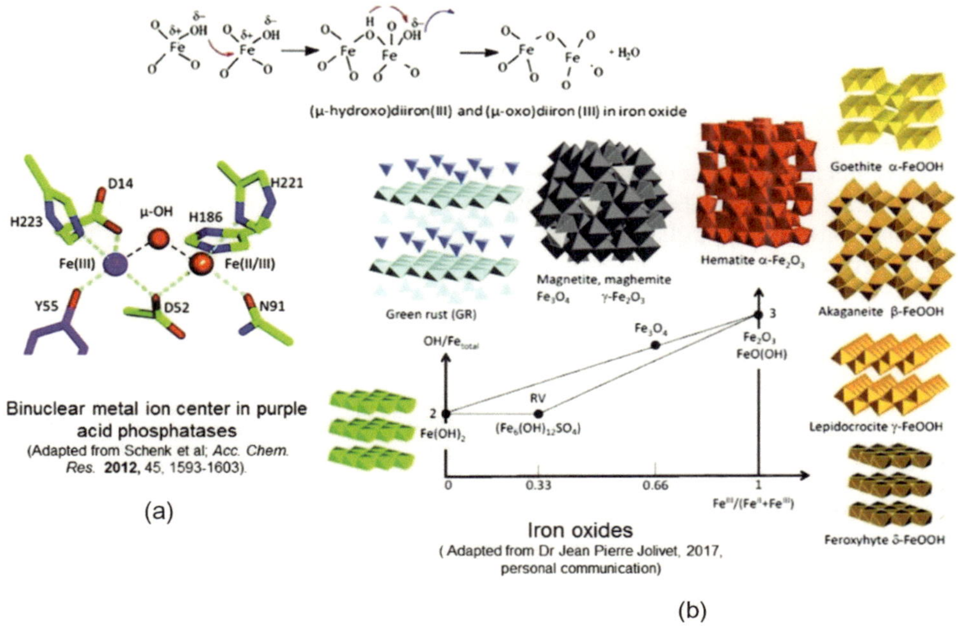

Figure 7.
The active form of μ-(hydr)oxo iron bridges in purple acid phosphatase (PAP) and different iron oxide phases [19]. Copyright 2018, Mary Ann Liebert, Inc.

natural phosphatase such as purple acid phosphatase (PAP). The authors highlighted that along with other studies of nanozymes such as iron oxide, vanadium pentoxide, and molybdenum trioxide, the oxo-metal bond in the oxide nanoparticles may play critical role for the catalysis in the corresponding natural metalloproteins. In particular, these inorganic nanoparticles with enzyme-like properties not only challenge the traditional concept of enzymes, but also provide new insights into life origin in the early Earth environments (**Figure 7**) [21, 22].

3. Iron oxide nanoparticles in biological system

Biogenic iron oxide nanoparticles, such as magnetosome and magnetoferritin, also perform enzyme-like property. Magnetosomes are often synthesized by magnetotactic bacteria species such as Alphaproteobacteria, Gammaproteobacteria, Deltaproteobacteria, and Nitrospira classes and the candidate phyla Latescibacteria (also known as candidate division WS3) and Omnitrophica (also known as candidate division OP3) of the Planctomycetes–Verrucomicrobia–Chlamydiae (PVC) bacterial superphylum [23]. The biogenic iron oxide nanoparticles have the single-domain size range of 35–120 nm and are covered by bacterial membranes. It was reported that magnetosomes exhibited peroxidase-like activity [24].

Magnetoferritin is caged protein with 24 subunits made up of heavy-chain ferritin (HFn) and light-chain ferritin. Ferritin is spherical, with an outer diameter of 12 nm and interior cavity diameter of 8 nm, in which iron oxide nanoparticles can be formed. It has been found that ferritins are natural nanozymes that exhibit intrinsic enzyme-like activities (e.g., ferroxidase, peroxidase) [25].

4. Broad applications based on bioactivity of iron oxide nanoparticles

The enzyme-like properties significantly extend the application range of iron oxide nanoparticles. Most of the biomedical applications of iron oxide nanoparticles are developed based on their excellent magnetism, such as magnetic separation of proteins, nucleic acids, or cells, hyperthermia of tumor, targeted drug delivery, and MRI contrast. After the discovery of enzyme-like activities, iron oxide nanoparticles have been extensively applied in immune detection, biosensor, antitumor, antibacterial, antiviral, antioxidant, and immune regulation.

First, the peroxidase-like activity allows iron oxide nanoparticles to be applied as an HRP alternative in immunoassays or biosensors for *in vitro* or *in vivo* detections [26]. Since iron oxide nanoparticles drive colorimetric reaction of chromogenic substrates, such as TMB, DAB, or ABTS, they can be used in ELISA to replace HRP for signal amplification [6, 27]. In recent year, Yan group has developed a lateral flow test using iron oxide nanoparticles, which is called as nanozyme strip [28]. In this strip, iron oxide nanoparticles are used to replace colloid gold nanoparticles and amplify the signal by their peroxidase-like activity, which significantly improves the detection sensitivity for EBOLA, flu virus, or SARS-COV-2 virus [29]. In addition, iron oxide nanoparticles coupled with natural enzymes such as glucose oxidase can be used for glucose detection *via* colorimetric reaction or electrochemical detection [30].

Second, owing to oxidoreductase-like activities, iron oxide nanoparticles perform the ability of ROS regulation, which is applied in the treatments of antitumor, antibacterial, antiviral, antioxidant. The peroxidase-like activity boosts ROS generation, which allows iron oxide nanoparticles to be used to kill bacteria [31] or tumor cells [32]. In addition, iron oxide nanoparticles perform antiviral activity by inducing lipid peroxidation in enveloped viruses and subsequently disrupt integrity of virus [17]. Besides generating ROS, iron oxide nanoparticles also can scavenge ROS by utilizing their catalase-like or SOD-like activity. Such unique property can be used for antioxidant treatments in diminishing cytotoxicity [14, 16], ischemia reperfusion of brain [33] and heart [34], neurodegeneration [35]. Recent studies demonstrate that iron oxide nanoparticles can regulate immune system to suppress tumor growth [36] or act as catalytic adjuvant to improve the immune effects of viral vaccine [37].

Besides biomedical applications, iron oxide nanoparticles also can be used with potential in other fields such as environment treatment. By utilizing peroxidase-like activity, iron oxide nanoparticles can be used to detect or degrade the pollutants in environment. For instance, hydrogen peroxide in acid rain can be detected using iron oxide nanoparticles [38]. Pollutants in wastewater, such as phenol, can be degraded by iron oxide nanoparticles [39]. Overall, the enzyme-like activities endow iron oxide nanoparticles with multifunctional property and extend their applications in many important fields.

5. Conclusion

The enzyme-like activities of iron oxide nanoparticles are a unique property for such inorganic nanomaterial. The catalytic types and efficiency are correlated with the nanostructure of iron oxide nanoparticles. Iron oxide nanoparticles act as enzyme mimics of natural enzymes whose active centers are composed of iron as a key cofactor, which not only extend their potential applications, but also indicate that inorganic nanomaterials are not biological inert but active to interact with biological system. These findings may provide a clue for the origin of life from inorganic world to organic and biological world. Though the catalytic efficiency is typically lower than their natural counterparts, iron oxide nanoparticles have high stability and can be scaled up with low cost, thus having a great potential to be used as enzyme mimics (Nanozymes) in many fields.

Acknowledgements

This work was supported by the National Natural Science Foundation of China grant (81930050).

Conflict of interest

The authors declare no conflict of interest.

Author details

Lizeng Gao
CAS Engineering Laboratory of Nanozyme, Institute of Biophysics, Chinese Academy of Sciences, Beijing, China

*Address all correspondence to: gaolizeng@ibp.ac.cn

References

[1] Jolivet JP, Tronc E, Chaneac C. Iron oxides: From molecular clusters to solid. A nice example of chemical versatility. Comptes Rendus Geoscience. 2006;**338**(6-7):488-497

[2] Guo HB, Barnard AS. Naturally occurring iron oxide nanoparticles: Morphology, surface chemistry and environmental stability. Journal of Materials Chemistry A. 2013;**1**(1):27-42

[3] Cornell RM, Schwertmann U. The Iron Oxides: Structure, Properties, Reactions, Occurrences and Uses. Second ed. Weinheim, Germany: Wiley-VCH; 2003. DOI: 10.1002/3527602097

[4] Baumgartner J, Faivre D. Iron solubility, colloids and their impact on iron (oxyhydr)oxide formation from solution. Earth-Science Reviews. 2015;**150**:520-530

[5] Vindedahl AM et al. Organic matter and iron oxide nanoparticles: Aggregation, interactions, and reactivity. Environmental Science: Nano. 2016;**3**(3):494-505

[6] Gao L et al. Intrinsic peroxidase-like activity of ferromagnetic nanoparticles. Nature Nanotechnology. 2007;**2**(9):577

[7] Wei H, Wang E. Nanomaterials with enzyme-like characteristics (nanozymes): Next-generation artificial enzymes. Chemical Society Reviews. 2013;**42**(14):6060-6093

[8] Wu JJX et al. Nanomaterials with enzyme-like characteristics (nanozymes): Next-generation artificial enzymes (II). Chemical Society Reviews. 2019;**48**(4):1004-1076

[9] Gao LZ, Fan KL, Yan XY. Iron oxide nanozyme: A multifunctional enzyme mimetic for biomedical applications. Theranostics. 2017;**7**(13):3207-3227

[10] Shen XM et al. Density functional theory-based method to predict the activities of nanomaterials as peroxidase mimics. ACS Catalysis. 2020;**10**(21):12657-12665

[11] Mo W-c et al. Reversible inhibition of Iron oxide nanozyme by guanidine chloride. Frontiers in Chemistry. 2020;**8**:491

[12] Jiang B et al. Standardized assays for determining the catalytic activity and kinetics of peroxidase-like nanozymes. Nature Protocols. 2018;**13**(7):1506-1520

[13] Zhang Z et al. Molecular imprinting on inorganic nanozymes for hundred-fold enzyme specificity. Journal of the American Chemical Society. 2017;**139**(15):5412-5419

[14] Chen ZW et al. Dual enzyme-like activities of iron oxide nanoparticles and their implication for diminishing cytotoxicity. ACS Nano. 2012;**6**(5):4001-4012

[15] Zhang RF et al. Unveiling the active sites on ferrihydrite with apparent catalase-like activity for potentiating radiotherapy. Nano Today. 2021;**41**:101317

[16] Gu YH et al. Vitamin B-2 functionalized iron oxide nanozymes for mouth ulcer healing. Science China. Life Sciences. 2020;**63**(1):68-79

[17] Qin T et al. Catalytic inactivation of influenza virus by iron oxide nanozyme. Theranostics. 2019;**9**(23):6920-6935

[18] Guo SB, Guo L. Unraveling the multi-enzyme-like activities of Iron oxide nanozyme via a first-principles microkinetic study. Journal of Physical Chemistry C. 2019;**123**(50):30318-30334

[19] Huang XL. Hydrolysis of phosphate esters catalyzed by inorganic iron oxide nanoparticles acting as biocatalysts. Astrobiology. 2018;**18**(3):294-310

[20] Huang XL, Zhang JZ. Hydrolysis of glucose-6-phosphate in aged, acid-forced hydrolysed nanomolar inorganic iron solutions-an inorganic biocatalyst? RSC Advances. 2012;**2**(1):199-208

[21] Huang XL, Zhang JZ. Sediment-water exchange of dissolved organic phosphorus in Florida bay. In: CIMAS Annual Report 2007-2008, Cooperative Institute for Marine and Atmospheric Studies. Miami, FL: University of Miami; 2008. pp. 72-73

[22] Huang XL. Iron oxide nanoparticles: An inorganic phosphatase. In: Sinha I, Shukla M, editors. Nanocatalysts. IntechOpen; 2019. DOI: 10.5772/intechopen.82650

[23] Uebe R, Schuler D. Magnetosome biogenesis in magnetotactic bacteria. Nature Reviews Microbiology. 2016;**14**(10):621-637

[24] Li KF et al. Magnetosomes extracted from Magnetospirillum magneticum strain AMB-1 showed enhanced peroxidase-like activity under visible-light irradiation. Enzyme and Microbial Technology. 2015;**72**:72-78

[25] Jiang B et al. Ferritins as natural and artificial nanozymes for theranostics. Theranostics. 2020;**10**(2):687-706

[26] Si-Rong L, Yen-Chun H, Jia-Rui L, Er-Kang W, Hui WEI. Nanozymes in analytical chemistry: From in vitro detection to live bioassays. Progress in Biochemistry and Biophysics. 2018;**45**(**2**):129-147

[27] Gao LZ et al. Magnetite nanoparticle-linked immunosorbent assay. Journal of Physical Chemistry C. 2008;**112**(44):17357-17361

[28] Duan DM et al. Nanozyme-strip for rapid local diagnosis of Ebola. Biosensors & Bioelectronics. 2015;**74**:134-141

[29] Liu D et al. Nanozyme chemiluminescence paper test for rapid and sensitive detection of SARS-CoV-2 antigen. Biosensors & Bioelectronics. 2020;**173**:112817

[30] Wei H, Wang E. Fe3O4 magnetic nanoparticles as peroxidase mimetics and their applications in H2O2 and glucose detection. Analytical Chemistry. 2008;**80**(6):2250-2254

[31] Gao L et al. Nanocatalysts promote Streptococcus mutans biofilm matrix degradation and enhance bacterial killing to suppress dental caries in vivo. Biomaterials. 2016;**101**:272-284

[32] Wang PX et al. Peroxidase-like nanozymes induce a novel form of cell death and inhibit tumor growth in vivo. Advanced Functional Materials. 2020;**30**(21):2000647

[33] Yan BC et al. Dietary Fe_3O_4 nanozymes prevent the injury of neurons and blood–brain barrier integrity from cerebral ischemic stroke. ACS Biomaterials Science & Engineering. 2020;7(1):299-310

[34] Xiong F et al. Cardioprotective activity of iron oxide nanoparticles. Scientific Reports. 2015;**5**:8579

[35] Zhang Y et al. Dietary iron oxide nanoparticles delay aging and ameliorate neurodegeneration in Drosophila. Advanced Materials. 2016;**28**(7):1387-1393

[36] Zanganeh S et al. Iron oxide nanoparticles inhibit tumour growth by inducing pro-inflammatory macrophage polarization in tumour tissues. Nature Nanotechnology. 2016;**11**(11):986-994

[37] Qin T et al. Mucosal vaccination for influenza protection enhanced by catalytic immune-adjuvant. Advanced Science. 2020;**7**(18):2000771

[38] Zhuang J et al. A novel application of iron oxide nanoparticles for detection of hydrogen peroxide in acid rain. Materials Letters. 2008;**62**(24):3972-3974

[39] Zhang JB et al. Decomposing phenol by the hidden talent of ferromagnetic nanoparticles. Chemosphere. 2008;**73**(9): 1524-1528

Section 4

Application

Chapter 4

Iron Oxide Nanoparticles and Nano-Composites: An Efficient Tool for Cancer Theranostics

Jaison Darson and Mothilal Mohan

Abstract

In recent years, functional Iron oxides nanoparticles and nano-composites have gained a special traction in the field of nano-biomedicine, owing to their multifunctional capabilities that includes the inherent magnetic resonance imaging, magnetic bioseparation, cargo delivery and magnetic hyperthermia behavior. Interestingly, there are various forms of iron oxides available, with each form having their own specific characteristics. The different polymorphic forms of iron oxides are obtained through various synthetic routes and are usually surface modified to prevent their oxidation. The chapter shall encompass the synthesis and surface modification of Iron oxides nanoparticles, physicochemical properties, and theranostic application of the magnetic iron oxide nanoparticles in cancer. Also, the future directions of Iron oxide nanoparticles and nano-composites towards the achievement of clinically realizable nanoformulation for cancer theranostic applications were highlighted.

Keywords: iron oxide nanoparticles, functionalised nanoparticles, hyperthermia, MRI contrasting ability, tumor ablation, tumor environment, nano-carrier, clinical translation

1. Introduction

Over the past few decades, nanomaterials have gained a special interest and are being investigated widely for various applications, owing to their unique characteristics [1]. Until now, several nanoparticulate systems have been studied to demonstrate their potential to detect, diagnose and treat cancer effectively with high degree of specificity and affinity to target cells, in comparison to the other onco-therapeutic approaches [1–3]. Iron oxides are being investigated widely in the field of nano-biomedicine, owing to their greater degree of variability and versatility [4]. The multifunctional capabilities of iron oxide nanoparticles including tumor labelling, magnetic bioseparation, biological entities detection, transfections, *invivo* cell tracking, tissue repair, clinical diagnosis, targeted drug delivery, magnetic hyperthermia and altered drug pharmacokinetics can be achieved via surface modification and bioconjugation [5, 6]. Iron oxides are found in many forms, namely magnetite (Fe_3O_4), hematite (α-Fe_2O_3), maghemite (γ-Fe_2O_3), wustite (FeO), bixbyite type

(β-Fe_2O_3), ε-Fe_2O_3 [6, 7], hydroxides, oxide/hydroxides [7] and ζ-Fe_2O_3 [8]. Each polymorphic form of Iron oxides has their own specific features [4]. Among these various forms, magnetite and maghemite are widely studied and found to be promising due to their proven biocompability [6] and unique magnetic properties [9]. With recent progress in nano-platforms, majority of studies are focused on the development of magnetic systems with long circulation half-life, high specific absorption rate, low Curie temperature [Wei Wu] and shortened transverse relaxation times, i.e., T_2 and T^*_2 [9, 10].

In general, the magnetic behavior of nanomaterials is greatly influenced by various features, such as size, size distribution, shape, polymorphic form, surface modification and purity [11–13]. Iron oxide nanoparticles of various polymorphic forms are synthesized through various approaches, including co-precipitation [14, 15], sol-gel method [16], sol-gel cum reverse-micelle technique [16], thermal decomposition [17], sonochemical [18], microwave heating [19, 20], hydrothermal [21–23], microemulsion [24–26], green synthesis [27–28], bacterial synthesis [29], laser pyrolysis [30], and electrochemical synthesis [31]. For effective practical applications, the nanoparticles and nano-composites must be readily aqueous dispersible, stable and biocompatible with fascinating magnetic properties and interactive surfaces. The particle characteristics of Iron oxides such as size, shape and surface charges serves as a determining factor in achieving biological distribution and elimination [32]. Multiple studies evidence that the particle characteristics play an important role in the toxicity, i.e. smaller particles showed increased toxicity than larger particles [33–35].

2. Synthesis and surface functionalization of iron oxides

In the last two decades, several synthetic approaches have been followed to synthesize iron oxide nanoparticles with controlled size and morphology, biocompatibility and monodisperse nature [32, 36]. The schematic depiction of commonly employed chemical approaches is shown in **Figure 1**.

Of many approaches, co-precipitation is widely used for the synthesis of iron oxide due to their process simplicity [37]. At the same time, it is necessary to consider several factors such as precursor salts used, ratio of Fe^{2+}/Fe^{3+} ions, nature and type of surfactants used, reaction parameters (pH and ionic strength of media, reaction temperature, stirring rate, drop rate of salt/base solution) and chemical processes employed in the co-precipitation technique to achieve the iron oxide nanoparticles with desirable crystal structure and morphology, monodispersity, and attractive magnetic properties [32, 38]. Peternele et al. synthesized magnetite and maghemite nanoparticles by precipitating the mixture of chloride salts of Fe^{2+} and Fe^{3+} (1:2 ratio) using 1.5 M NaOH or 25% Ammonia under vigorous stirring. The sequences of addition of base strongly affect the formation of phase and size of the magnetite nanoparticles. Further, much smaller sized maghemite was obtained by oxidizing prepared magnetite nanoparticles due to the Ostwald ripening effect. The nanoparticles obtained using ammonia displayed uniform size and were monodisperse in nature [37]. Cui et al. synthesized magnetite, maghemite and hematite using a common epoxide precipitation route involving an ethanolic solution of ferrous chloride tetrahydrate and a gelation agent propylene oxide. Initially, the nucleation of magnetite was observed due to the oxidation of Fe(II) precursor. From this step, the magnetite solution can be transformed into Fe_3O_4, γ-Fe_2O_3 and α-Fe_2O_3 nanoparticles through centrifugation-followed by vacuum drying, sol-to-xerogel formation followed

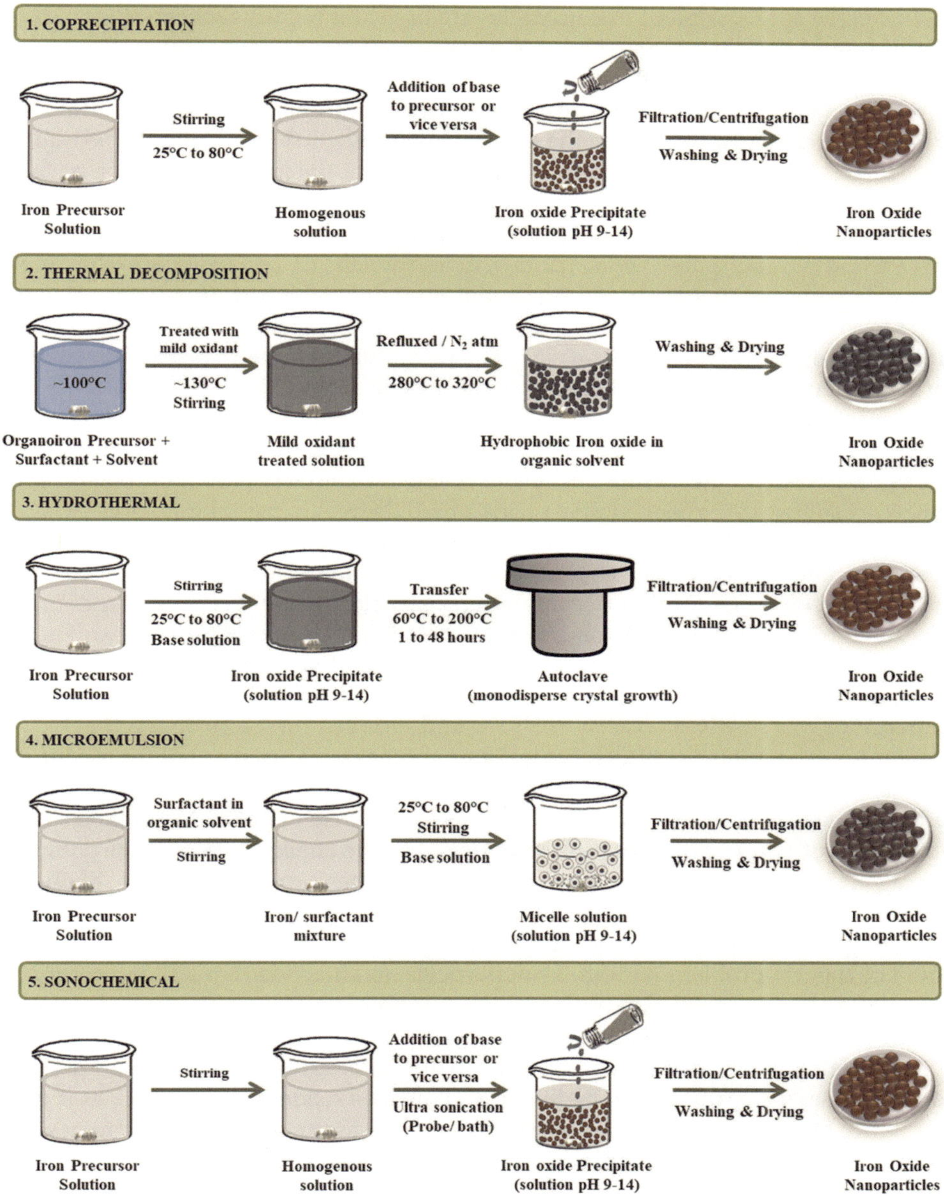

Figure 1.
Schematic depiction of commonly used chemical techniques.

by oxidation at 150°C and, atmosphere controlled evaporation of sol at 150°C, respectively [39, 40]. In another study, Cui et al. synthesized nanoparticles of other polymorphic forms such as lepidocrocite and goethite just by regulating the pH of the obtained greenish precipitate comprising an ethanolic solution of ferrous chloride tetrahydrate and a gelation agent propylene oxide using ammonia, followed by air oxidation under room temperature [41].

In several studies, the microemulsion and reverse micellization has been employed for the preparation of iron oxide nanoparticles with controlled morphology, but at the same time the aggregation of nanoparticles are high and requires several washing, and stabilization [32]. For instance, Lee et al. synthesized magnetite nanoparticles by adding the hydro-ethanolic Fe (II, III) salt precursor solution into sodium dodecylbenzenesulfonate (SDBS) dispersed xylene solution under vigorous stirring for 12 h. Hydrazine was added to the solution maintained at 90°C, and the solution was refluxed for 5 h to obtain nanoparticles. The size of the nanoparticles can be tuned by modifying the relative amounts of precursor and the polar solvent-surfactant ratio [42]. In another study, Jung et al. applied the reverse micelle technique to synthesize uniform and monodisperse Fe_3O_4 nanoparticles for sensing and drug targeting applications. Herein, the precursor and surfactant immobilized in the organic phase were added to the aqueous phase comprising stabilizer and stirred for several hours to form stable reverse micelles of sub 3 nm size. The formation of large irregular and small worm-like nanoparticles was observed in the absence of stabilizer [43].

In order to achieve iron oxide nanoparticles with high-quality monodisperse nature, thermal decomposition approach has been used widely. However, this requires relatively higher temperature and involves complex operations. Hyeon et al. prepared maghemite nanocrystallites with a high degree of crystallinity and monodiperse nature by subjecting Iron pentacarbonyl-oleic acid complex to thermal decomposition. Initially, the complexation was carried out by transferring the Iron pentacarbonyl into a hot mixture containing oleic acid and octyl ether and maintained at 100°C for 1 h. The resulting mixture was cooled and treated with mild oxidant trimethylamine oxide. Following the addition of trimethylamine oxide, the solution was again heated to 130°C for 2 h under an inert (Ar) atmosphere. Further, the temperature was slowly increased to 300°C and refluxed for 1 h to obtain the maghemite nanocrystallites [44]. Later, Lassenberger et al. used a slightly modified thermal decomposition technique to synthesize the monodisperse oleic acid stabilized Iron oxide nanoparticles [45], which does not involve the use of mild oxidants. The particle size of nanocrystallites is directly linked to the concentration of the complexation agent and the heating rate employed in the synthesis process [44, 45]. Zhou et al. successfully synthesized various morphologies of monodisperse Fe_3O_4 nanoparticles by varying the ratio of Iron oleate/sodium oleate, reflux temperature, and heating rate [46]. The biggest problem with the nanoparticles obtained via thermal decomposition route is their limited solubility in aqueous environments. Thus, phase transformation is required to render them water soluble [45].

A simple hydrothermal process can be a better alternative for the preparation of monodisperse, dislocation-free, highly crystalline iron oxide nanoparticles, as it does not require high temperature. The use of surfactants in the hydrothermal process could limit the growth of nanoparticle size, while retaining the crystallinity and magnetic properties close to that of the iron oxides prepared without surfactants and bulk iron oxides [6, 32]. Ozel et al. synthesized Iron oxides of varying crystallinity and size ranging from 12 to 49 nm by varying the reaction temperature (60–180°C) and reaction time (1–48 h) [47]. Further, Torres-Gomez et al. reported the synthesis of various γ-Fe_2O_3 nanostructures such as quasi-spherical, octahedral and truncated cubes by varying the reaction temperature from 120 to 160°C [22]. Ellipsoid 3D superstructures, plate-like nanostructures and irregular structures of α-Fe_2O_3 were obtained by employing varying proportions of Fe^{3+} precursor, surfactants and solvent under varying temperatures [48]. Xu et al. used urea to synthesize template-free rod-like β-FeOOH structures by varying the reaction temperature and time [49]. Also, urea in combination with ammonia was added to ferric chloride solution and autoclaved at 150°C for 6 h to obtain β-Fe_2O_3 [50]. Various reports highlight that depending on the type of the reducing agent and surfactants used in the hydrothermal process, different iron oxide phases and nanostructures such as spherical, polyhedral, nanocubes, octahedral, truncated cubes, hollow spheres, nanorods etc. [6, 22, 23, 32, 47, 49, 51, 52] can be obtained.

Sonochemical technique is a competitive alternative, to prepare ultrafine, monodispersive iron oxide nanoparticles with unusual properties. Usually, iron oxides prepared through this route are amorphous and possess low magnetization with speromagnetic behavior below magnetic transition temperature [6]. Hassanjani-Roshan et al. demonstrated the effect of ultrasonic power and reaction temperature on the particle characteristics such as crystalline/amorphous behavior and particle size of α-Fe_2O_3. The particle can be transformed into a highly crystalline form by subjecting them to higher temperature, following the sonochemical synthesis [53]. A cube-like Fe_3O_4 nanoparticle with different particle sizes ranging from 20 to 58 nm was obtained by varying the molar concentration of the precursor and the reducing agent [54]. According to the LaMer model, the local supersaturation should be higher to achieve the smaller particles. Ludtke-Buzug et al. state that the maximum local supersaturation is higher at lower ultrasound frequency [55].

Recently, green chemistry and biological methods are being used for the preparation of iron oxide nanoparticles owing to their safety, low cost, non-toxic and eco friendliness approach. However, the consistency of synthetic process is highly influenced by the source of the green and biological reducing agents [40]. The microbial-mediated approach demands a lengthy incubation time for the synthesis of iron oxide nanoparticles [56]. Whilst, the plant-mediated environment-friendly approach requires less time and can be used for the synthesis of various iron oxide nanostructures like spherical, needles, cubical, dendrites, cylindrical, and so on [56] with appreciable biological activities. Many studies utilize plant extracts with mild reducing capability along with bases like sodium hydroxide [57, 58], sodium carbonate [59], glycine [60] to synthesize stabilized iron oxide nanoparticles, as this can avoid the use of environmentally-toxic stabilizing agents. Few studies reported the successful synthesis of Iron oxide nanostructures using the extracts like grape berry ferment [61], flower extract of *Avecinnia marina* [62], and leaf extract of *Bauhinia tomentosa* [27] alone. The synthesis techniques of Iron oxide nanoparticles with their characteristic size are depicted in Table A1.

Avoiding agglomeration while retaining the stability is the most crucial challenge every magnetic material undergoes. In last two decades, considerable efforts have been devoted for the passivation of iron oxide nanoparticle surface using organic and inorganic materials, to avoid agglomeration, and to have improved solubility, stability and biocompatibility. The surface functionalization of iron oxide nanoparticles could be achieved by coating the iron oxide core with shell material (or) by dispersing the iron oxide core over the shell material (or) through chemical interaction of iron oxides with shell material (or) through bioconjugation reactions (or) by coating a shell-core structure with shell material. The shell materials could be organic or inorganic materials with functional properties. The commonly employed functional materials in the passivation of iron oxide surface include organic small molecules (drugs, fatty acids, polyol, dyes, vitamins, cyclodextrins), surfactants (dextran, polyvinyl alcohol), biological molecules (proteins, nucleic acids, antibody, cells, enzymes, microbes), silanes (*p*-aminophenyltrimethoxysilane, 3-aminopropyltriethyloxysilane), polymers (synthetic and natural origin), silica, metals and non-metals (gold, silver, carbon, etc.), metal oxides and metal sulphides [6]. In recent times, with the advancement in the polymer technology, several stimulus-induced targeted iron oxide-drug nanoconjugates are developed. Inspired by the different pH conditions of tumor environment, several pH-sensitive systems are being explored to minimize the release of chemotherapeutic drugs in the blood and normal tissues [61, 63–66]. Also, the thermo-responsive systems are being explored simultaneously to have control over the drug release, i.e., the release of drug is initiated only at the hyperthermia temperature [66, 67–70]. Interestingly, other stimulus-mediated systems such as enzyme-mediated [71, 72], light-mediated [71, 73], ultrasound-mediated [71, 74] are also being explored.

3. Physicochemical properties of iron oxides

3.1 Crystallographic properties and polymorphism

Iron (III) oxides exhibit distinct physical properties due to its polymorphic forms that has same molecular formula but with different crystallographic structures. Iron (III) oxides occur in various polymorphic forms such as α-Fe_2O_3, β-Fe_2O_3, γ-Fe_2O_3, ε-Fe_2O_3, and ζ-Fe_2O_3. The polymorphic forms are characterized by crystal structure as: (i) rhombohedrally-centric hexagonal α-Fe_2O_3 ($R\bar{3}c$); (ii) body-centric cubic β-Fe_2O_3 ($Ia\bar{3}$); (iii) inverse spinel type cubic γ-Fe_2O_3 ($Fd\bar{3}m$) with oxygen vacancies in octahedral site; (iv) orthorhombic ε-Fe_2O_3 ($Pna2_1$); and (v) monoclinic ζ-Fe_2O_3 (I2/a). All polymorphic form of Iron (III) oxides can be converted into other phases in response to temperature and pressure. The transformations of phases are interdependent on the precursors, pH value, ionic strength, intrinsic properties of nanomaterials such as crystal structure, particle characteristics, matrix characteristics in which particles are incorporated, reaction parameters and the nature of the treatment applied (e.g. thermal/pressure/both) [6, 16, 75]. However, under ambient conditions, a thermodynamically stable α-Fe_2O_3 phase is usually formed directly from β-Fe_2O_3, ε-Fe_2O_3, and γ-Fe_2O_3. Temperature treatment plays a major role in the formation of α-Fe_2O_3 phase from the other phases such as β-Fe_2O_3 ($\sim$773 K) [76], γ-Fe_2O_3 ($\sim$973 K) [77] and ε-Fe_2O_3 (>1473 K) [78] phases. In some cases, like hollow β-Fe_2O_3, the phase can be transformed directly into γ-Fe_2O_3, and depending on the interparticle interactions and size of the nanoparticles, γ-Fe_2O_3 may exhibit both direct and indirect transformations (via ε-Fe_2O_3) into α-Fe_2O_3. Recently, a stable ζ-Fe_2O_3 was formed by exposing β-Fe_2O_3 above 30 GPa pressure [75, 76]. The other predominantly investigated polymorphic form includes the face-centric cubic Fe_3O_4 ($FeO.Fe_2O_3$, II and III oxidation states) with tetrahedral and octahedral sites. Like other Iron oxides, Iron (II, III) oxides also tend to undergo phase transformation to other Iron (III) oxide polymorphs under ambient conditions, due to their poor stability in oxygen environment [6]. It is well accepted that the crystalline phase stability of nanoparticles are much dependent on the surface stress, surface strain and surface energy. Recently, much focus has been kept on Cubic Wustite (FeO, space group Fm3m), which under ambient conditions remains in metastable state and undergo oxidation to other polymorphic form of iron oxides [6, 79–80]. Like iron oxides, iron (III) oxyhydroxides also exists in various forms namely an orthorhombic goethite (α-FeOOH, space group Pnma), monoclinic akaganeite (β-FeOOH, space group I2/m), orthorhombic lepidocrocite (γ-FeOOH, space group Cmcm) and hexagonal feroxyhyte (δ-FeOOH) [81]. Ferrihydrite are another polymorphic form of iron (II) oxides that are relatively abundant in the natural systems and can be readily transformed into the goethite phase when there is a rapid oxidation process [82]. Among various iron (III) oxide polymorphs, ε-Fe_2O_3 particles are observed only in the nanosize and they are size dependent [83]. An amorphous form of Fe_2O_3 was observed in the particles with less than 5 nm in diameter, wherein the oxygen octahedra are randomly oriented around the Fe (III) ions. However, it is difficult to distinguish the ultrafine particles of amorphous iron oxides and other polymorphs experimentally [84].

3.2 Magnetic properties

Magnetic properties of the nanomaterials play a key role in MRI contrasting ability, magnetically-induced heating, externally-targeted drug delivery and bio-sensing applications. The γ-Fe_2O_3 polymorph exhibits ferrimagnetic and superparamagnetic behavior with a curie temperature of 928 K [16, 76], whilst, the ferromagnetic ε-Fe_2O_3 possess highest coercivity [76, 84] with a curie temperature close to 500 K [85]. Interestingly, the magnetic order of ε-Fe_2O_3 nanoparticles does not get vanished even at 500 K and this different ferromagnetic state persists up to 850 K [86]. A weak ferromagnetic behavior was reported in α-Fe_2O_3 phase with a Morin transition at ~269 K, i.e., a transition of antiferromagnetic state from weak ferromagnetic state [76] and a curie temperature of 950 K [16]. The paramagnetic β-Fe_2O_3 becomes magnetically ordered below the Neel temperature (~110 K) and exhibit antiferromagnetic state. Similarly, the recently identified ζ-Fe_2O_3 exhibited an antiferromagnetic nature below Neel transition temperature of ~69 K [76]. A paramagnetic transition of amorphous Fe_2O_3 at temperatures above the Neel temperature of 80 K was concluded based on the interpretation on Mössbauer data [84]. In some cases, non-ideal magnetic behavior could be observed in iron oxides due to wide range of blocking temperatures, aroused from wide range of particle size distribution [6]. Wustite (FeO) is generally stable above 560°C and possess antiferromagnetic nature with a Neel temperature of about 200 K [79]. Among all Iron (III) oxyhydroxide, δ-FeOOH is the only polymorphic form which showed significant magnetization at room temperature with ferrimagnetic behavior (Tc—450 K) [87] and significant relaxation properties [88]. The smaller crystals of Feroxyhites (δ-FeOOH) showed speromagnetic behavior between 80 K and 300 K [87]. β-FeOOH is usually paramagnetic at room temperatures and below Neel temperature they exhibit antiferromagnetic property [89]. The bulk α-FeOOH and γ-FeOOH displayed a Neel transition at 252 K and 53 K respectively [90]. The magnetic property of Ferrihydrite is size dependant, which displayed antiferromagnetic behavior below 4 nm, and above 4 nm ferrimagnetic behavior was observed [91]. The magnetic properties of iron oxide nanoparticles are greatly influenced by the oxidation and aggregation. The oxidation of iron oxide nanoparticles could lead to the loss of magnetic properties. In contrast the aggregation of particles may lead to mutual magnetisation that is usually aroused by the interaction of magnetic field of one nanoparticle with the neighboring nanoparticle [32].

3.3 Chemical properties

Iron oxide nanoparticles are highly prone for oxidation, particularly in the atmospheric air, and hence require a thin and non-interactive protective coating that has minimal effect on its characteristic physical properties, especially its magnetic properties [32]. Many studies also concluded that the naked iron oxide nanoparticles tend to agglomerate owing to their high surface energy, surface area-to-volume ratio, magnetic interactions and van der Waals forces [32, 92]. The agglomeration of the particles not only increases the particle size, but also enhances strong magnetic dipole-dipole attractions, that make the particles ferromagnetic. In general, the hydrophobic character and the huge surface area-to-volume ratio render the iron oxide nanoparticles toxic, insisting the need for the modification of particle surface, to make them hydrophilic and biocompatible [32].

4. Theranostic applications of magnetic iron oxide nanoparticles in cancer

4.1 Tumor imaging

Among various imaging modalities, magnetic resonance imaging acquires a rapid image of *invivo* tumor environment owing to their high spatial resolution. The unique magnetic properties with shorten relaxation times of iron oxide nanoparticles could enhance the sensitivity of T2 and T2* contrasting ability in magnetic resonance imaging (MRI) [9]. The imaging sensitivity and specificity of iron oxide nanoparticles could be enhanced by modifying the surface of iron oxides with small molecules, peptides and antibodies, whose receptors are overexpressed on the surface of tumor cells. Yang et al., modified the poly(amino acid)-coated iron oxide nanoparticle surface with Her-2/neu antibody for the detection of HER2/neu positive breast cancer cells and observed an enhanced MRI contrasting ability with significant T_2 relaxation time. Over many years, a well-known antibody herceptin is used clinically against the HER2/neu receptor [93]. Lee et al., demonstrated the detection of small tumors using the herceptin conjugated iron oxide nanoparticles [94]. Though antibodies show promising future for onco-imaging and therapy, their steric effects limits their conjugation with iron oxides. Further, the decrease in specificity of antibody-iron oxide conjugate was observed due to the interaction of Fc receptors of normal tissues with the antibody. Recently, single chain antibodies (scFv) have gained more traction owing to their small molecular weight, antigen binding ability, non-immunogenicity and low cost. Some scFv's such as scFvEGFR10 and sm3E have been conjugated to the iron oxide nanoparticles for the reduction of T_2 relaxation time and enhanced MRI capability. Yang and his co-workers conjugated amino-terminal fragment (ATF) peptide and near infrared dye Cy5.5 into the iron oxide-amphiphillic polymer conjugate to target the urokinase plasminogen activator receptor that is over expressed in human tumor cells and stromal cells where the tumor is associated. Herein, the study reported that the imaging probe is capable to detect small tumors ranging from 0.5 to 1 mm^3. The other peptides that are generally used to achieve multi-modal imaging capability include $EPPT_1$ peptide, Luteinizing hormone release hormone, Folic acid, Vitamin FA and Arg-Gly-Asp [95–97] and chlorotoxin peptide [98]. The iron oxides could also possess T_1 relaxation, when the particle size is greatly reduced, i.e. less than 5 nm. This phenomenon is observed due to the decrease in magnetic moment exerted with decreased particle size. The Iron (II, III) oxides below 5 nm are considered to be responsive MRI contrast agents due to its ability to be used as both T_1 and T_2 contrasting ability. Under tumor environment, the iron oxides are released from the carrier providing the T_2-T_1 switching ability during imaging [95–97]. Despite of the establishment of *in vitro* imaging potential of tumor targeted iron oxides, they are not widely used in clinical practice owing to their low specificity and sensitivity *in vivo*. Also, for successful clinical use, the targeted iron oxides must be able to monitor tumor metastasis and therapy response. The low sensitivity of single imaging modality has evolved the multimodal imaging technique which involves the combination of MRI contrast imaging with photothermal imaging [95].

4.2 Cancer labelling and sorting

Cell labelling and sorting is a key technique in the field of oncology and stem cell research. Usually, cells are labeled using ferro or superpara or paramagnetic materials in the process of *invivo* cell separation, and the labeled cells are visualized using the magnetic resonance imaging based protocol. In principle, two approaches are employed for cell labelling using iron oxides; (i) one by directly attaching the magnetic iron oxide nanoparticles to the cell, and (ii) the other by internalizing the iron oxide nanoparticles into the cytosol of cell using receptor-mediated endocytosis or fluid phase endocytosis or phagocytosis [99–101]. Cell labelling must not compromise the proliferating capability, motility, and cellular functions for effective stem cell trials or therapies. Arbab et al. labeled CD34+ hematopoietic and mesenchymal stem cells with ferrumoxides-protamine sulfate complex and reported the unaltered expression of phenotypic markers and differentiating capacity [102]. The cell viability remained intact when labeling concentration of iron oxide nanoparticles was used. In contrast, at high concentrations, the cellular viability decreased. The uptake of iron oxide into stem cells can be enhanced by modifying the core with dextran, citrates, HIV tat peptide, unfractionated heparin, and aminosilane. The modification of the iron oxide core exerts various biological activities in labeled cells. For instance, there is an increased cell proliferation in iron oxide-ferucarbotran labeled mesenchymal stem cells. The target-specific molecules/Iron oxide nanoconjugate can precisely label the cells *in vivo* [103]. Chen et al. demonstrated the MRI imaging potential of Herceptin bound-dextran-coated iron oxides in HER2/neu receptors expressed cancer cell lines [104]. In another study, the contrasting ability of folic acid bound-PEG coated iron oxides in the tumor cells expressing folate receptors [105].

The magnetic behavior of iron oxides allows the isolation of specific cells from the biological suspensions like blood, apart from its diagnostic ability. In non-invasive magnetic-activated cell sorting, the antibody-bound iron oxides bind specifically to the specific antigens present on the surface of target cells. The bound fraction can be separated from the unbound fraction by applying a magnetic field. For instance, circulating tumor cells are captured, and analyzed for staging cancer, selection of therapy and monitoring treatment [106]. Mi et al. developed a low cytotoxic Herceptin-functionalised conjugated oligomer-based Iron oxide-silica nanoparticle system for magnetic-activated sorting and fluorescence-activated detection of circulating tumor cells at its metastatic stage [107]. Du et al. bioengineered D-tyrosine phosphate decorated iron oxides that can be dephosphorylated to tyrosine coated iron oxides by the overexpressed alkaline phosphatases in the surface of cancer cells. The tyrosine-coated iron oxides upon dephosphorylation have been attached to the tumor cells and captured using a small magnet [108].

4.3 Cancer immunotherapy

Immunotherapy is a robust strategy to eliminate cancer cells. For effective cancer immunotherapy, the immune system must be activated against tumor antigens. The immunotherapeutic approaches include cell-based immunotherapy, monoclonal antibodies (mAb) for checkpoint blockade and cancer vaccines. In cell-based immunotherapy, the migration, expansion, and depletion of immune cells are tracked to understand the complex cellular and molecular mechanism involved in the immune system [100]. The schematic representation of various applications of iron oxide nanoparticles is depicted in **Figure 2**.

4.3.1 Cell-based immunotherapy

The T cells have the ability to differentiate into various forms based on the interaction of specific tumor antigens on the antigen-presenting cells. Real-time T-cell trafficking using MRI can improve the T cell-based immunotherapies by achieving better localization [109]. For instance, the three-dimensional MRI visualization of *in vivo* T-cell trafficking to target tumors and the temporal regulation of T-cells within the tumors has been demonstrated using ova-specific CD8+ T cells labeled cross-linked iron oxides (highly derivatized). In addition, a similar proliferating potential, cellular interaction, and cytotoxic profiles was observed for both labeled and unlabelled cells [110]. Liu et al. developed PEG conjugated fluorescent dyes coated iron oxide for achieving dual-mode imaging such as MRI and Fluorescence. The

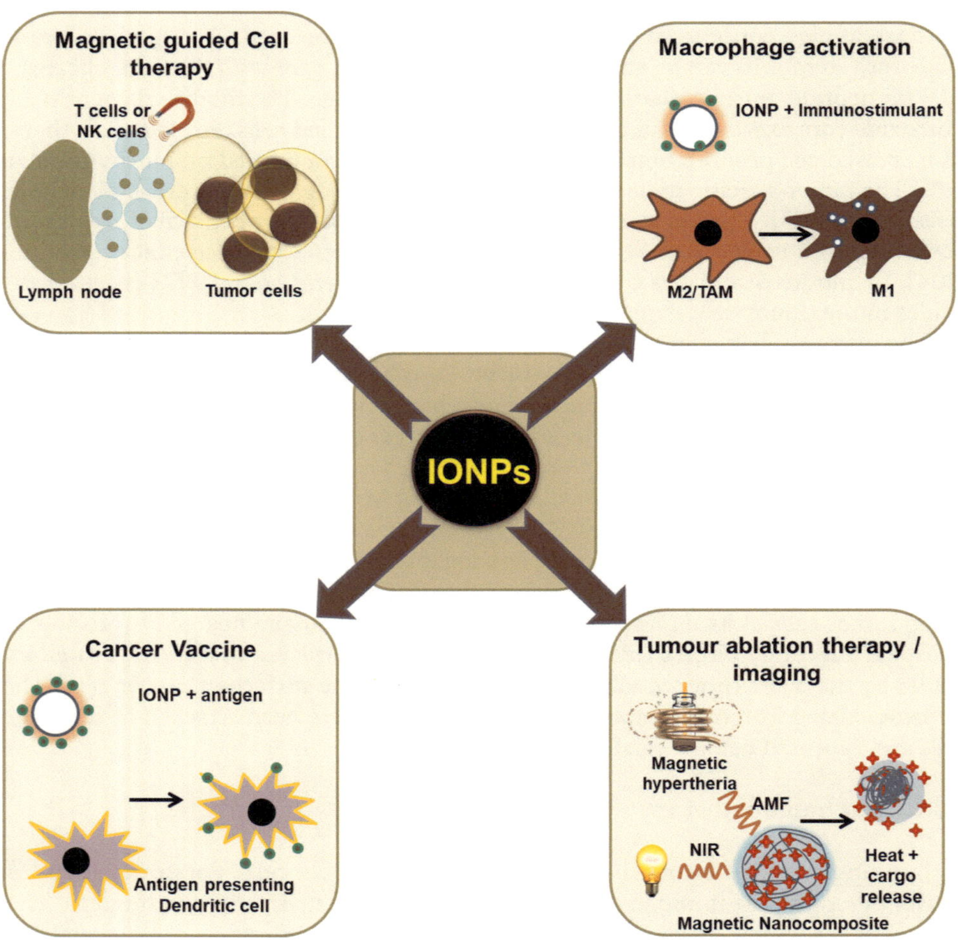

Figure 2.
Application of iron oxide nanoparticles/conjugates in cancer immunotherapy.

human and murine T cellular functions were not altered following the injection of these nanoparticles [111]. The labeling of T cells is the biggest challenge due to their non-phagocytic nature. However, significant internalization of Iron oxides can be achieved using the transfection agents such as polyethyleneimine, poly-L-lysine, lipofectamine [100].

Macrophages eliminate foreign particles and cellular wastes by secreting cytokines and initiating phagocytosis. Reprogramming or polarizing the tumor-associated macrophages (TAM) in the tumor environment shall overcome the difficulty of penetration of M1 macrophages from the outside environment. Iron oxides on their own activate the macrophages through metabolic degradation. In a study, the response to CD47 monoclonal antibodies by tumor-associated macrophages (TAM) has been monitored using ferumoxytol-enhanced MRI. The study showed that phagocytosis of TAM had been activated due to the inhibition of the interaction between SIRPα and CD47 by CD47 mAb [112]. In another study, 3-methyladenine was incorporated into hollow iron oxide nanoparticles to promote the immune response by reprograming the TAM into M1-type macrophages. The nanoparticle system upregulates the NF-κB p65 while inhibiting the expression of P13Kγ to promote an immune response. The mouse model revealed a synergistic effect of polarization of macrophages by 3-methyladenine and iron oxide nanoparticles [113]. An artificial reprogramming of macrophages was reported in a study using hyaluronic acid-coated iron oxide nanoparticles. In this study, the iron oxide nanoparticles and macrophages from RAW 264.7 mouse were incubated together and injected into 4 T1 tumor-bearing mice. The results indicated the capability of these artificially programmed macrophages in polarization of TAM and enhanced tumor effect [114].

Dendritic cells play a key role in activating cytotoxic T lymphocytes and regulating adaptive immune response by presenting the tumor antigens into the draining lymph nodes. The immune response is highly dependant on the migration of activated dendritic cells to lymph nodes [103, 115]. The incorporation of iron oxides into a dendritic cell does not impair the viability of cells [103] and demonstrate the tracking of dendritic cells with MRI [100]. The antigens are loaded into immature dendritic cells, as the phagocytic ability is higher than the mature dendritic cells. The internalization of iron oxides takes place via endocytosis for the particles ranging from 20 to 200 nm [100]. Iron oxide nanoparticles bound to oval albumin showed increased expression of TNFα, IL-6, and IFN-γ in murine dendritic cells. Further the assessment of immunotherapeutic capability in mice revealed a dramatic reduction in the tumor [116, 117]. De Vries et al. reported the enhanced accuracy of magnetic resonance tracking of iron oxides-loaded dendritic cells over scintigraphic imaging. Herein, the iron oxide nanoparticles are phagocytized into the autologous *ex vivo* cultured dendritic cells without altering the phenotype and functional properties [115]. Dendritic cells loaded with iron oxide/zinc oxide core-shell nanoparticles exhibited a T_2-weighted signal reduction in the lymph node of C57BL/6 mice. However, the dendritic cells loaded with zinc oxide showed a migration towards the lymph nodes [100].

4.3.2 Cancer vaccines

Cancer vaccines play a crucial role in presenting the tumor antigens to activate specific T cells against the tumor cells. Recently, biomimetic nano vaccines encompassing tumor cell membranes with tumor antigens are used for targeting the immune or tumor cells. The vaccination capacity can be increased by introducing iron oxides with intrinsic magnetic properties. For instance, Zhang et al. developed a biomimetic magnetosome to expand and stimulate antigen-specific cytotoxic T-cell and to track and effectively guide them into tumor tissues [118]. Further, Wang et al., demonstrated antitumor activity and immunogenic cell death with low systemic toxicity using Ce6-loaded magnetic/mesoporous organosilica nanoparticles concealed with cancer cells [119]. In another study, Long CM et al. developed magneto-vaccination using iron oxide labeled GM-CSF secreting cells mixed with tumor to stimulate the immune system by inducing the T cell tumor targeting factors proliferation [120].

4.3.3 Checkpoint blockade

The immune checkpoint is the molecular interactions between cancer cells and immune cells. The ability of tumor cells to evade the surveillance of the immune system is the common problem associated with the T cell-mediated approach, as many healthy cells like certain tumor cells encompass inhibitory checkpoint programmed cell death protein 1 ligand 1 (PD-L1) that can inactivate the T cells by binding to the inhibitory checkpoint PD1 protein expressed on T cell surface [100]. The immune response shall be enhanced by targeting inhibitory checkpoint molecules such as proteins and antibodies. Under physiological conditions, the checkpoint inhibitory molecules are highly prone to degradation and hence required to be encapsulated into a robust delivery system. In a study, the immunoswitch design was demonstrated using antibodies-loaded dextran-coated iron oxide nanoparticles for the inhibition of immune checkpoints. The antibodies against PD-L1 and 4-1BB were used to stimulate T cells. In this study, the tumor-bearing mice (C57BL/6) was initially treated with adoptively transferred T cells, followed by the administration of antibodies-loaded dextran-coated iron oxide nanoparticles or free individual antibodies. The results suggest the targeting of multiple checkpoints, as targeting only one checkpoint did not result in size reduction of tumors. Also, highlights the multifunctional utility of iron oxides for checkpoint inhibition [121]. In another study, the Iron oxide-coated folic acid-functionalised-disulphide-polyethylene glycol-conjugated polyethylenimine nanoparticles were developed for the delivery of siRNA to inhibit PD-L1 protein. The nanoparticles exhibited higher transfection ability, MRI contrasting ability, and the high cellular uptake downregulate the PD-L1, which in turn affected the T-cells cytokine-secretion level [122]. The schematic representation of immune checkpoint regulation using immunoswitch nanoparticles is shown in **Figure 3**.

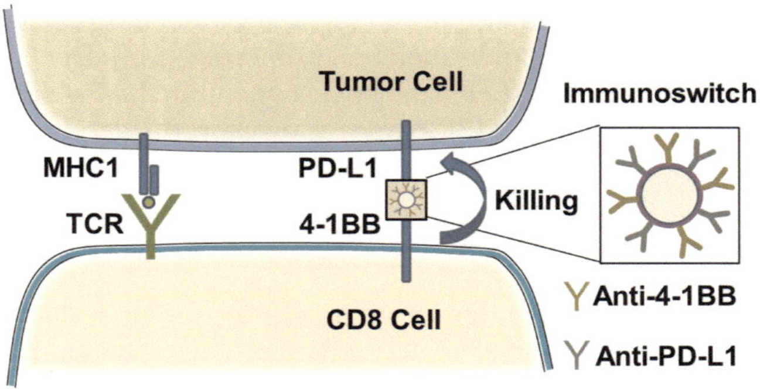

Figure 3.
Schematic representation of immune check point regulation using immunoswitch nanoparticle.

4.4 Tumor ablation therapy

In tumor ablation therapy, the use of non-contact magnetic heating (above 42°C) has gained a special attention and are being utilized in clinical practice in few hospitals as an adjuvant therapy. The magnetic heating is achieved through hysteresis loss and relaxation losses under varying alternating magnetic field and radio frequencies. The efficiency of magnetic heating is directly linked to the size, shape and concentration of the magnetic nanoparticles, the strength and frequency of alternating magnetic field and cooling rate in biological tissues. The major obstacle in the use of Iron oxides for non-contact heating of tumor tissues is their low heating power. Magnetite and maghemite nanoparticles are widely employed for magnetic hyperthermia applications due to their well-established biocompatibility. However, these nanoparticles possess relatively low coercivity and require high applied frequency, usually ranging from 400 kHz to 900 kHz, to effectively heat the media in which it is dispersed. In contrast, studies indicate that ε-Fe_2O_3 nanoparticles exhibit hyperthermia potential at low applied frequency of about 20 kHz to 100 kHz. Though many magnetic materials reported to have high heating power, their concern over safety has limited their use in clinical practice, and encourages the optimisation of structural features of iron oxide nanoparticles for enhanced clinical hyperthermia potential [123].

Kolosnjaj-Tabi et al. demonstrated a mild hyperthermia efficacy of PEG-coated iron oxide nanocubes in a magnetic field of 23.8 kA/m and 111 kHz until the particle resides in the interstitial extracellular location. The hyperthermia efficiency of the nanoparticles was lost after cellular internalization and capture in the liver and spleen. However, the hyperthermia effect destabilize the tumor stroma to enhance the drug penetration [124]. The cancer theranostic agents are widely employed for the effective control of tumor owing to their potential diagnostic cum treatment approach. The magnetic hyperthermia efficiency and MRI T_2 contrasting ability of iron oxide nanoparticles were demonstrated using the fourth-generation dendrimer coated iron oxides [125]. Hayashi et al., developed iron oxide nanoclusters for combined MRI cum hyperthermia, as the individual iron oxides (<10 nm) are prone to leakage from capillaries. In this study, the surface of Iron oxide nanoclusters is modified with polyethylene glycol and folic acid to enhance their accumulation within the tumor environment of mice, following the intravenous administration. The mice that underwent local heating for 20 minutes reduced the tumor volume to about 1/10th of the control mice, indicating the hyperthermia efficiency of the iron oxide nanoclusters [126]. Lin et al. developed a multifunctional pegylated albumin nanocomplex comprising Iron oxide and a hydrophobic dye (IR780). The photothermal effect and MR imaging of nanocomplex were demonstrated on a cancer colon model and tumor-bearing mice, respectively [127]. The combined photothermal effect (NIR-induced) and cancer imaging (MRI and fluorescence) were demonstrated using a novel dumbbell-like Gold-Iron oxide nanoparticle by Kirui et al. [128]. A similar kind of bimodal cancer imaging with a photothermal effect was reported in hyaluronan-targeted iron oxides to bring out the photothermal efficiency and cell staining potential of the iron oxides apart from its MRI contrasting ability [129]. Espinosa et al. demonstrated a complete tumor regression using the dual-mode hyperthermia and photothermal therapeutic potential of iron oxide nanocubes [130].

4.5 Drug delivery applications

The successful delivery of therapeutic agents into the tumor environment with minimal toxicity to surrounding tissues is the biggest challenge, as it is often limited by tumor heterogeneity, dense fibrotic stromal barriers and various vascular barriers including abnormal tumor blood vessels, tumor cells proliferating nests, normal blood vessels, positive intratumoural pressure [131, 132]. Iron oxide nanocarrier systems are often believed to overcome these biological barriers by altering the pharmacokinetics and tissue distribution profile via enhanced permeability and retention effect [9]. However, for clinical success, EPR effect may not be sufficiently enough as it offers only certain level of tumor targeting and non-specific biodistribution. Thus, the logical choice for tumor therapy shall be actively tumor-targeted iron-oxide nano-carrier system, which can be minimally toxic to normal tissues while having enhanced bioavailability, intracellular bio-distribution, and potent cyto-toxic effects against the tumor cells. [131, 132]. Recently, the tumor-targeted iron oxide nanoparticles are used to monitor the accumulation of drugs in the tumor site, while simultaneously estimating the drug level in the tumor tissues. The detection of MRI signal changes of drug loaded iron oxides can provide a track over drug delivery, estimated drug levels in tissue and therapeutic response *invivo* [9].

Li et al. developed the iron oxide nanoclusters with photothermal mediated synergistic chemotherapy and chemodynamic therapy. In this study, the core iron oxides were surface modified using the paclitaxel-loaded human serum albumin and conjugated to the Arg-Gly-Asp peptides for tumor specific targeting [133]. Several studies indicate that insufficient penetration through BBB reduces the efficiency in treating glioblastoma multiforme. Norouzi developed pH-sensitive doxorubicin-loaded Iron oxide nanoparticles stabilized with trimethoxysilylpropyl-ethylenediamine triacetic acid to demonstrate their uptake in brain-derived using mouse model. The cellular uptake of nanoparticles was increased by 2.8 folds and provided an enhanced anti-tumor effect than free doxorubicin. Further, the study indicated that the penetration of nanoparticles into the brain was augmented due to the combination of cadherin binding peptide and external alternating magnetic field [134]. Hussein-Al-Ali et al. developed iron oxide/chlorambucil/chitosan nanocomposite with a particle size of about 15 nm. Controlled release of drug chlorambucil from the nanocomposite with significant anti-tumor effects on leukemia cancer cell lines was observed [135].

5. Concluding remarks and future directions of iron oxides

Since several decades, many studies have been conducted to evaluate the potential use of functionalised iron oxide nanoparticles for the delivery of anticancer drugs, yet there are several obstacles that need to be overcome for increased adoption of these nano-carrier systems into clinical practice. The various challenges that needs to be considered while developing the newer targeted iron oxide nanoparticles includes the synthesis of conjugated iron oxide nanoparticles without inducing a change in the chemical and magnetic properties, the high drug loading efficiency, regulation of circulation time [113], specificity and selectivity of towards tumor tissues, and controlled release of drug with the tumor region. The other challenges that are least explored in the development of iron oxide nanocarrier systems includes tumor uptake, biodistribution and bioelimination. The biological distribution is highly dependent on the nanoparticles size, morphology and surface characteristics, as these properties can strongly influence the particle interaction with serum proteins and cells [32]. The major obstacle in achieving the effective tumor therapy is the tumor heterogeneity resulted from the genetic mutations. This emphasis the need for personalized medicine involving both imaging and targeted drug delivery simultaneously, signifying the concept 'we observe what we treat'. The potent MRI capability of the iron oxide nanoparticles shall allow the visualization of events such as delivery of drug and other cargo molecules, efficacy of the undergoing treatment, gene expression and metastasis, bioelimination of iron oxides. Though, iron oxides exhibit many distinctive properties, the long-term fate of iron oxides, PK/PD studies, toxicology studies and the toxicity criteria are yet to be clearly defined. Recently, there is an intense focus on the development of multifunctional tumor-targeted drug-loaded iron oxide nano-carrier systems, as it can offer many benefits such as, tumor-specific targeting, MRI contrasting ability (track and monitor the accumulation of the nano-carrier system), combined hyperthermia and chemotherapy to tumor cells and stimulus-induced drug release (control over the drug release at the tumor-specific site). Until today, there are numerous studies reported to have such multifunctional properties in *in vitro* and *in vivo* animal models. The successful development of clinically reliable, multifunctional tumor-targeted drug-loaded Iron oxide nano-carrier systems can transform the future oncology treatment practices [32].

Appendix

Nanoparticle type	Precursor	Precipitating agent	Process and conditions	Size	Ref.
Co-precipitation technique					
Fe_3O_4	$FeCl_2·4H_2O$, $FeCl_3·6H_2O$	1.5 M NaOH	Addition of precursor solution into alkali,		[37]
			a. stirring under air for 30 min	9.82 nm	
			b. N_2 bubbling (100 mL/min) for 30 min	6.31 nm	
			c. kept under magnetic field for 24 h	7.21 nm	
Fe_3O_4	$FeCl_2·4H_2O$, $FeCl_3·6H_2O$	25% NH_4OH	Addition of alkali into precursor solution		[37]
			a. stirring under air for 30 min	11.22 nm	
			b. N_2 bubbling (100 mL/min) for 30 min	5.05 nm	
			c. kept under magnetic field for 24 h	5.41 nm	
γ-Fe_2O_3	$FeCl_2·4H_2O$, $FeCl_3·6H_2O$	1.5 M NaOH	Addition of precursor solution into alkali, adjusted to pH 3.5 using HCl		[37]
			a. stirring under air for 1 h at 95°C	7.21 nm	
			b. O_2 flux (100 mL/min) for 1 h at 95°C	7.21 nm	
			c. Oxidation at 250°C for 3 h	7.21 nm	
γ-Fe_2O_3	$FeCl_2·4H_2O$, $FeCl_3·6H_2O$	25% NH_4OH	Addition of alkali into precursor solution, adjusted to pH 3.5 using HCl		[37]
			a. stirring under air for 1 h at 95°C	6.31 nm	

Nanoparticle type	Precursor	Precipitating agent	Process and conditions	Size	Ref.
			b. O_2 flux (100 mL/min) for 1 h at 95°C	5.41 nm	
			c. Oxidation at 250°C for 3 h	5.41 nm	
Fe_3O_4	$FeCl_2{\cdot}4H_2O$	Propylene oxide	78°C for 30 min, centrifugation and vacuum drying	7.5 nm	[40]
γ-Fe_2O_3	$FeCl_2{\cdot}4H_2O$	Propylene oxide	78°C for 30 min, xerogel formation (100°C), followed by oxidation at 150°C	8.6 nm	[40]
α-Fe_2O_3	$FeCl_2{\cdot}4H_2O$	Propylene oxide	78°C for 30 min, atmosphere controlled evaporation at 150°C	18.4 nm	[40]
γ-FeOOH	$FeCl_2{\cdot}4H_2O$	Propylene oxide	60°C for 30 min, pH 6.5, air oxidation for 5 h	40–80 nm	[41]
α-FeOOH	$FeCl_2{\cdot}4H_2O$	Propylene oxide Ammonia	60°C for 30 min, pH 8 to 8.5, air oxidation for 20 h	20–90 nm	[41]
Fe_3O_4	$FeCl_2{\cdot}4H_2O$	Propylene oxide Ammonia	60°C for 30 min, pH ≥9, air oxidation for 2 h	25–50 nm	[41]
Fe_3O_4	$Fe(SO_4){\cdot}7H_2O$, $Fe(NO_3)_3{\cdot}9H_2O$	NaOH	Primary nucleation was carried out by adding NaOH solution into precursor solution at 60°C under stirring and after 10 min HCl was added to the precipitate and stirred for 1 h	7.8 nm	[11]
Reverse micelle and microemulsion technique					
Fe_3O_4	Fe(III) stearate	hydrazine	Organic phase—precursor, olyelamine and xylene; aqueous phase—F127 in water; reverse micelles solution heated to 90°C, hydrazine addition and stirring for 3 h	2.8 nm	[43]
Fe_3O_4	Fe(III) oleate	hydrazine	Organic phase—precursor, olyelamine and xylene; aqueous phase—F127 in water; reverse micelles solution heated to 90°C, hydrazine addition and stirring for 3 h	7 nm	[43]
Fe_3O_4	Fe(III) acetylacetonate Oleic acid	hydrazine	Organic phase—precursor, olyelamine and xylene; aqueous phase—F127 in water; reverse micelles solution heated to 90°C, hydrazine addition and stirring for 3 h	<2 nm	[43]

Nanoparticle type	Precursor	Precipitating agent	Process and conditions	Size	Ref.
Fe_3O_4	$FeCl_2{\cdot}4H_2O$, $Fe(NO_3)_3{\cdot}9H_2O$	hydrazine	Organic phase—SDBS, xylene; aqueous phase-ethanol, precursor salts in water; reverse micelles solution heated to 90°C, hydrazine addition and refluxed for 5 h	2–10 nm	[42]
Thermal decomposition technique					
γ-Fe_2O_3	$Fe(CO)_5$	Trimethalamine oxide	Complexation—precursor, octylether, oleic acid; maintained at 100°C for 1 h; treated with trimethyloxide; heated to 130°C under Ar atmosphere; refluxed at 300°C	4–16 nm	[44]
Iron oxide	$Fe(CO)_5$	—	Complexation—precursor, dioctylether, oleic acid; raised to 100°C, N_2 atmosphere, and held for 10 min; refluxed at 290°C for 1 h	3–10 nm	[45]
Fe_3O_4	Iron oleate	Isopropanol	Precursor, sodium oleate (different concentration), oleic acid dissolved in 1-octadecene or tri-n-octylamine; refluxed at 300 to 360°C (with varying heating rate from 5 to 15°C/min) for 2 to 6 h	Diverse morphology and size	[46]
Hydrothermal technique					
Iron oxide	$FeCl_3{\cdot}6H_2O$, $FeCl_2{\cdot}4H_2O$	Ammonia	The primary nanoparticles prepared by adding ammonia into precursor solution were grown in an autoclave at 60°C (12 h), 100°C (12 h, 72 h), 150°C (12 h, 24 h), 180°C (1 h, 12 h, 24 h, 48 h)	12–49 nm	[47]
γ-Fe_2O_3	$FeCl_3{\cdot}6H_2O$	Hydrazine	The primary nanoparticles prepared by adding hydrazine into precursor/PEG8000 solution were grown in an autoclave at 120°C, 140°C, 160°C for 4 h	Diverse morphology and size	[Torrez]
α-Fe_2O_3	$FeCl_3{\cdot}6H_2O$	Ammonium hydroxide	The solution containing different precursor, solvent and additives was autoclaved at 160°C/180°C for 10–12 h	Diverse morphology and size	[48]

Nanoparticle type	Precursor	Precipitating agent	Process and conditions	Size	Ref.
β-FeOOH rods	$FeCl_3 \cdot 6H_2O$	Urea	70–110°C for 8 h	0.4–2.3 μm	[49]
β-Fe_2O_3	$FeCl_3 \cdot 6H_2O$	Urea, Ammonia	pH 9.66, 150°C for 8 h	50–90 nm	[50]
Sonochemical technique					
α-Fe_2O_3	$FeCl_3 \cdot 6H_2O$	NaOH	The precursor solution was added dropwise into NaOH solution and subjected to varying sonication power (1, 3, 5, 7, 9 W) and reaction temperature for 30 min; annealed for 1 h	Amorphous/ crystalline particles	[53]
Green chemistry and biological methods					
Fe_2O_3 rods	$Fe(SO_4) \cdot 7H_2O$	Leaf extract/ NaOH	To the precursor solution, a mixture of *Lantana camara* leaf extract and NaOH was added; precipitate obtained was air-dried	10–20 nm	[57]
Fe_2O_3	$FeCl_3 \cdot 6H_2O$	Leaf extract	To the Bauhinia tomentosa leaf extract, the precursor solution was added; precipitate obtained was air-dried	70 nm	[27]
Iron oxide	$FeSO_4 \cdot 7H_2O$	Henna extract/ NaOH	Henna extract was added to precursor solution maintained at 60 °C for 30 min; NaOH is added to form Iron oxides	150–200 nm	[58]
Iron oxide	$FeCl_3 \cdot 6H_2O$	Leaf extract/ Na_2CO_3	One part of precursor solution was mixed with leaf extract of *Cymbopogon citratus* and heated to 60°C for 1 h; two parts of precursor added to the mixture and temperature was raised to 85°C; pH raised to 10 using Na_2CO_3 to precipitate Iron oxide	9 ± 4 nm	[59]
Fe_3O_4	$FeCl_2 \cdot 4H_2O$, $FeCl_3 \cdot 6H_2O$	Plant extract/ glycine	Hot precursor solution was added to whole plant extract of *Kappaphycus alvarezii* and Glycine; stirred for 60 min; heated upto 300°C for 15 min; annealed in air at 500°C for 2 h	10–30 nm	[60]
Fe-O	$FeCl_3 \cdot 6H_2O$	Flower extract	Precursor solution was added dropwise into flower extract of *Avecinnia marina*; centrifuged; dried under vacuum at 125°C for 2 h	10–40 nm	[61]

Nanoparticle type	Precursor	Precipitating agent	Process and conditions	Size	Ref.
γ-Fe_2O_3	Fe $(NO_3)_3 \cdot 9H_2O$	Grape berry ferment	The pH of the precursor solution was adjusted from 2.7 to 1.5 using grape berry ferment; transferred to autoclave and kept at 200° C for 24 h; iron oxide precipitate was obtained	6–18 nm	[62]

Table A1.
Synthesis techniques of iron oxide nanoparticles with its characteristic size.

Author details

Jaison Darson* and Mothilal Mohan
Department of Pharmaceutics, SRM College of Pharmacy, Kattankulathur, India

*Address all correspondence to: jaipharma17@gmail.com

References

[1] Gharpure KM, Wu SY, Li C, Berestein GL, Sood AK. Nanotechnology: Future of oncotherapy. Clinical Cancer Research. 2015;**21**(14):3121-3130

[2] Hartshorn CM, Bradbury MS, Lanze GM, Nel AE, Rao J, Wang AZ, et al. Nanotechnology strategies to advance outcomes in clinical cancer care. ACS Nano. 2018;**12**(1):24-43

[3] Madamsetty VS, Mukherjee A, Mukherjee S. Recent trends of the bio-inspired nanoparticles in cancer theranostics. Frontiers in Pharmacology. 2019;**10**:1264. DOI: 10.3389/fphar.2019.01264

[4] Soenen SJ, Cuyper MD, Smedt SCD, Braeckmans K. Investigating the toxic effects of Iron oxide nanoparticles. Methods in Enzymology. 2012;**509**: 195-224. DOI: 10.1016/B978-0-12-391858-1.00011-3

[5] Farzin A, Etesami SA, Quint J, Memic A, Tamayol A. Magnetic nanoparticles in cancer therapy and diagnosis. Advanced Healthcare Materials. 2020;**9**(9):1901058

[6] Wu W, He Q, Jiang C. Magnetic iron oxide nanoparticles: Synthesis and surface functionalization strategies. Nanoscale Research Letters. 2008;**3**:397

[7] Rahman MM, Khan SB, Jamal A, Faisal M, Aisiri AM. Iron Oxide Nanoparticles. In: Nanomaterials. London, UK: IntechOpen Limited; 2011. pp. 43-65. DOI: 10.5772/27698. [Chapter 3]

[8] Tucek J, Machala L, Ono S, Namai A, Yoshikiyo M, Imoto K, et al. Zeta-Fe_2O_3—A new stable polymorph in iron(III) oxide family. Scientific Reports. 2015;**5**:15091

[9] Peng X, Chen H, Huang J, Mao H, Shin DM. Targeted magnetic iron oxide nanoparticles for tumour imaging and therapy. In: Rezai RF, editor. Biomedical Engineering – From theory to applications. London, UK: IntechOpen; 2011. DOI: 10.5772/22873

[10] Sun C, Lee JSH, Zhang M. Magnetic nanoparticles in MR imaging and drug delivery. Advanced Drug Delivery Reviews. 2008;**60**(11):1252-1265

[11] Janani V, Induja S, Jaison D, Abhinav EM, Mothilal M, Gopalakrishnan C. Tailoring the hyperthermia potential of magnetite nanoparticles via gadolinium ion substitution. Ceramics International. 2021;**47**(22):31399-31046. DOI: 10.1016/j.ceramint.2021.08.015

[12] Jeevanandam J, Barhoum A, Chan YS, Dufresne A, Danquah MK. Review on nanoparticles and nanostructured materials: History, sources, toxicity and regulations. Beilstein Journal of Nanotechnology. 2018;**9**:1050-1074

[13] Guisbiers G, Mejía-Rosales S, Deepak FL. Nanomaterial properties: Size and shape dependencies. Journal of Nanomaterials. 2012;**2012**: Article ID 180976

[14] Besenhard MO, LaGrow AP, Hodzic A, Kriechbaum M, Panariello L, et al. Co-precipitation synthesis of stable iron oxide nanoparticles with NaOH: New insights and continuous production via flow chemistry. Chemical Engineering Journal. 2020;**399**:125740

[15] Yaid NA, Joon YC. Co-precipitation synthesis of magnetic nanoparticles for

efficient removal of heavy metal from synthetic wastewater. AIP Conference Proceedings. 2019;**2124**:020019

[16] Sakurai S, Namai A, Hashimoto K, Ohkoshi S. First observation of phase transformation of all four Fe_2O_3 phases ($\gamma \rightarrow \varepsilon \rightarrow \beta \rightarrow \alpha$-phase). Journal of the American Chemical Society. 2009;**131**: 18299-18303

[17] Unni M, Uhl AM, Savliwala S, Savitzky BH, Dhavalikar R, Garraud N, et al. Thermal decomposition synthesis of iron oxide nanoparticles with diminished magnetic dead layer by controlled addition of oxygen. ACS Nano. 2017;**11**(2):2284-2303

[18] Dolores R, Raquel S, Adianez G-L. Sonochemical synthesis of iron oxide nanoparticles loaded with folate and cisplatin: Effect of ultrasonic frequency. Ultrasonics Sonnochemistry. 2015;**23**: 391-398

[19] Aivazoglou E, Metaxa E, Hristoforou E. Microwave-assisted synthesis of iron oxide nanoparticles in biocompatible organic environment. AIP Advances. 2018;**8**(4):048201

[20] Hu L, Percheron A, Chaumont D, et al. Microwave-assisted one-step hydrothermal synthesis of pure iron oxide nanoparticles: Magnetite, maghemite and hematite. Journal of Sol-Gel Science and Technology. 2011;**60**: 198

[21] Ge S, Shi X, Sun K, Li C, Uher C, Baker JR, et al. Facile hydrothermal synthesis of iron oxide nanoparticles with tunable magnetic properties. The Journal of Physical Chemistry C. 2009; **113**(31):13593-13599

[22] Torres-Gómez N, Nava O, Argueta-Figueroa L, García-Contreras R, Baeza-Barrera A, Vilchis-Nestor AR. Shape tuning of magnetite nanoparticles obtained by hydrothermal synthesis: Effect of temperature. Journal of Nanomaterials. 2019;**2019**:7921273

[23] Ankamwar B. Hydrothermal synthesis of rod shaped iron oxide nanoparticles. AIP Conference Proceedings. 2021;**2335**(1):080002. DOI: 10.1063/5.0046421

[24] Kekalo K, Koo K, Zeitchick E, Baker I. Microemulsion synthesis of iron core/iron oxide shell magnetic nanoparticles and their physicochemical properties. Materials Research Society Symposium Proceedings. 2012;**1416**. DOI: 10.1557/opl.2012.736

[25] Housaindokht MR, Nakhaei Pour A. Size control of iron oxide nanoparticles using reverse microemulsion method: Morphology, reduction, and catalytic activity in CO hydrogenation. Journal of Chemistry. 2013;**2013**:781595

[26] Rose LC, Suhaimi H, Mamat M, Lik TZ. Synthesis and characterization of oleic acid surface modified magnetic iron oxide nanoparticles by using biocompatible w/o microemulsion for heavy metal removal. AIP Conference Proceedings. 2017;**1885**(1):020113. DOI: 10.1063/1.5002307

[27] Lakshminarayanan S, Shereen MF, Niraimathi KL, et al. One-pot green synthesis of iron oxide nanoparticles from *Bauhinia tomentosa*: Characterization and application towards synthesis of 1, 3 diolein. Scientific Reports. 2021;**11**:8643

[28] Bhuiyan MSH, Miah MY, Paul SC, Aka TD, Saha O, Rahaman MM, et al. Green synthesis of iron oxide nanoparticle using Carica papaya leaf extract: Application for photocatalytic degradation of remazol yellow RR dye

and antibacterial activity. Heliyon. 2020; **6**(8):e04603

[29] Majeed S, Danish M, Mohamad Ibrahim MN, et al. Bacteria mediated synthesis of iron oxide nanoparticles and their antibacterial, antioxidant, cytocompatibility properties. Journal of Cluster Science. 2021;**32**:1083-1094

[30] Greculeasa SG, Palade P, Schinteie G, et al. Tuning structural and magnetic properties of Fe oxide nanoparticles by specific hydrogenation treatments. Scientific Reports. 2020;**10**: 17174

[31] Ramimoghadam D, Bagheri S, Hamid SBA. Progress in electrochemical synthesis of magnetic iron oxide nanoparticles. Journal of Magnetism and Magnetic Materials. 2014;**368**: 207-229

[32] Ali A, Zafar H, Zia M, Haq IU, Phull AR, Ali JS, et al. Synthesis, characterisation, applications, and challenges of iron oxide nanoparticles. Nanotechnology, Science and Applications. 2016;**9**:49-67

[33] Gilmour MI, O'Connor S, Dick CA, Miller CA, Linak WP. Differential pulmonary inflammation and in vitro cytotoxicity of size-fractionated fly ash particles from pulverized coal combustion. Journal of the Air & Waste Management Association. 2004;**54**: 286-295

[34] Oberdorster G. Pulmonary effects of inhaled ultrafine particles. International Archives of Occupational and Environmental Health. 2001;**74**:1-8

[35] Vogel CFA, Charrier JG, Wu D, McFall AS, Li W, Abid A, et al. Physicochemical properties of iron oxide nanoparticles that contribute to cellular ROS-dependent signaling and acellular production of hydroxyl radical. Free Radical Research. 2016;**50**(11):1153-1164

[36] Khan I, Saeed K, Khan I. Nanoparticles: Properties, applications and toxicities. Arabian Journal of Chemistry. 2019;**12**(7):908-931. DOI: 10.1016/j.arabjc.2017.05.011

[37] Peternele WS, Monge Fuentes V, Fascineli ML, Rodrigues da Silva J, Silva RC, Lucci CM, et al. Experimental investigation of the coprecipitation method: An approach to obtain magnetite and maghemite nanoparticles with improved properties. Journal of Nanomaterials. 2014;**2014**:682985

[38] Aragaw TA, Bogale FM, Aragaw BA. Iron-based nanoparticles in wastewater treatment: A review on synthesis methods, applications, and removal mechanisms. Journal of Saudi Chemical Society. 2021;**25**(8):101280

[39] Chang M, Shih YH. Synthesis and application of magnetic iron oxide nanoparticles on the removal of Reactive Black 5: Reaction mechanism, temperature, and pH effects. Journal of Environmental Management. 2018; **224**(2018):235-242

[40] Cui H, Liu Y, Ren W. Structure switch between α-Fe_2O_3, γ-Fe_2O_3 and Fe_3O_4 during the large scale and low-temperature sol-gel synthesis of nearly monodispersed iron oxide nanoparticles. Advanced Powder Technology. 2013;**24**: 93-97

[41] Cui H, Ren W, Lin P, Liu Y. Structure control synthesis of iron oxide polymorph nanoparticles through an epoxide precipitation route. Journal of Experimental Nanoscience. 2013;**8**(7–8):869-875

[42] Lee Y, Lee J, Bae CJ, Park J-G, Noh H-J, Park J-H, et al. Large-scale synthesis

of uniform and crystalline magnetite nanoparticles using reverse micelles as nanoreactors under reflux conditions. Advanced Functional Materials. 2005; **15**(3):503-509. DOI: 10.1002/adfm.200400187

[43] Jung E, Kim SW, Cho A, Kim YJ, Jeong GJ, Kim J, et al. Synthesis of sub 3 nm-sized uniform magnetite nanoparticles using reverse micelle method for biomedical application. Materials (Basel). 2019;**12**(23):38-50

[44] Hyeon T, Lee SS, Park J, Chung Y, Na HB. Synthesis of highly crystalline and monodisperse maghemite nanocrystallites without a size-selection process. Journal of the American Chemical Society. 2001; **123**(51):12798-12801

[45] Lassenberger A, Grünewald TA, van Oostrum PDJ, Rennhofer H, Amenitsch H, Zirbs R, et al. Monodisperse iron oxide nanoparticles by thermal decomposition: elucidating particle formation by second-resolved in situ small-angle X-ray scattering. Chemistry of Materials. 2017; **29**(10):4511-4522

[46] Zhou Z, Zhu X, Wu D, Chen Q, Huang D, Sun, et al. Anisotropic shaped iron oxide nanostructures: controlled synthesis and proton relaxation shortening effects. Chemistry of Materials. 2015;**27**(9):3505-3515

[47] Ozel F, Kockar H, Karaagac O. Growth of iron oxide nanoparticles by hydrothermal process: Effect of reaction parameters on the nanoparticle size. Journal of Superconductivity and Novel Magnetism. 2014;**28**(3):823-829

[48] Tadic M, Trpkov D, Kopanja L, Vojnovic S, Panjan M. Hydrothermal synthesis of hematite (α-Fe_2O_3) nanoparticle forms: Synthesis conditions, structure, particle shape analysis, cytotoxicity and magnetic properties. Journal of Alloys and Compounds. 2019;**792**:599-609

[49] Xu Z, Liang J, Zhou L. Template-free hydrothermal synthesis of β-FeOOH nanorods and their catalytic activity in the degradation of methyl orange by a photo-Fenton-like process. OJINM. 2013;**3**(4):58-65

[50] Rahman MM, Jamal A, Khan SB, Faisal M. Characterization and applications of as-grown β-Fe_2O_3 nanoparticles prepared by hydrothermal method. Journal of Nanoparticle Research. 2011;**13**(9):3789-3799

[51] Jung HS, Choi HJ. Hydrothermal fabrication of octahedral-shaped Fe_3O_4 nanoparticles and their magnetorheological response. Journal of Applied Physics. 2015;**117**(17):17E708

[52] Yang Y, Huang M, Qian J, et al. Tunable Fe_3O_4 nanorods for enhanced magnetic hyperthermia performance. Scientific Reports. 2020;**10**:8331

[53] Hassanjani-Roshan A, Vaezi MR, Shokuhfar A, Rajabali Z. Synthesis of iron oxide nanoparticles via sonochemical method and their characterization. Particuology. 2011;**9**(1):95-99

[54] Fuentes-Garcia JA, Alavarse AC, Maldonado ACM, Toro-Cordova A, Ibarra MR, Goya F. Simple sonochemical method to optimize the heating efficiency of magnetic nanoparticles for magnetic fluid hyperthermia. ACS Omega. 2020;**5**:26357-26364

[55] Lüdtke-Buzug K, Penxová Z. Superparamagnetic iron oxide nanoparticles: An evaluation of the sonochemical synthesis process. Current Directions in Biomedical Engineering. 2019;**5**(1):307-309. DOI: 10.1515/cdbme-2019-0077

[56] Priya N, Kaur K, Sidhu AK. Green synthesis: An eco-friendly route for the synthesis of iron oxide nanoparticles. Frontiers in Nanotechnology. 2021;**3**: 655062

[57] Rajiv P, Bavadharani B, Kumar MN, Vanathi P. Synthesis and characterization of biogenic iron oxide nanoparticles using green chemistry approach and evaluating their biological activities. Biocatalysis and Agricultural Biotechnology. 2017;**12**:45-49. DOI: 10.1016/j.bcab.2017.08.015

[58] Chauhan S, LSB U. Biosynthesis of iron oxide nanoparticles using plant derivatives of Lawsonia inermis (Henna) and its surface modification for biomedical application. Nanotechnology for Environmental Engineering. 2019; **4**(1):8. DOI: 10.1007/s41204-019-0055-5

[59] Patino-Ruiz D, Sanchez-Botero L, Tejeda-Benitez L, Hinestroza J, Herrera A. Green synthesis of iron oxide nanoparticles using *Cymbopogon citratus* extract and sodium carbonate salt: Nanotoxicological considerations for potential environmental applications. Environmental Nanotechnology Monitoring & Management. 2020;**14**: 100377

[60] Arularasu MV, Devakumar J, Rajendran TV. An innovative approach for green synthesis of iron oxide nanoparticles: Characterization and its photocatalytic activity. Polyhedron. 2018;**156**:279-290

[61] Jaison D, Gopalakrishnan C, Mothilal M. pH-sensitive natural almond gum hydrocolloid based magnetic nanocomposites for theragnostic applications. International Journal of Biological Macromolecules. 2020;**154**: 256-266

[62] Karpagavinayagam P, Vedhi C. Green synthesis of iron oxide nanoparticles using Avicennia marina flower extract. Vacuum. 2018;**160**:286-292

[63] Zhu L, Wang D, Wei X, Zhu X, Li J, Tu C, et al. Multifunctional pH-sensitive superparamagnetic iron-oxide nanocomposites for targeted drug delivery and MR imaging. Journal of Controlled Release. 2013;**169**(3):228-238

[64] Lu J, Sun J, Li F, Wang J, Liu J, Kim D, et al. Highly sensitive diagnosis of small hepatocellular carcinoma using pH-responsive iron oxide nanocluster assemblies. Journal of the American Chemical Society. 2018;**140**(32): 10071-10074

[65] Espinola-Portilla F, Serrano-Torres O, Hurtado-López GF, Sierra U, Varenne A, d'Orlyé F, et al. Superparamagnetic iron oxide nanoparticles functionalized with a binary alkoxysilane array and poly(4-vinylpyridine) for magnetic targeting and pH-responsive release of doxorubicin. New Journal of Chemistry. 2021;**45**(7):3600-3609

[66] Amin M, Huang W, Seynhaeve ALB, ten Hagen TLM. Hyperthermia and temperature-sensitive nanomaterials for spatiotemporal drug delivery to solid tumors. Pharmaceutics. 2020;**12**(11): 1007

[67] Kalele S, Narain R, Krishnan KM. Probing temperature-sensitive behavior of pNIPAAm-coated iron oxide nanoparticles using frequency-dependent magnetic measurements. Journal of Magnetism and Magnetic Materials. 2009;**321**(10):1377-1380

[68] Tai LA, Tsai PJ, Wang YC, Wang YJ, Lo LW, Yang CS. Thermosensitive liposomes entrapping iron oxide nanoparticles for controllable drug release. Nanotechnology. 2009;**20**(13): 135101

[69] Mai BT, Balakrishnan PB, Barthel MJ, Piccardi F, Niculaes D, Marinaro F, et al. Thermoresponsive iron oxide nanocubes for an effective clinical translation of magnetic hyperthermia and heat-mediated chemotherapy. ACS Applied Materials & Interfaces. 2019; **11**(6):5727-5739

[70] Hemery G, Garanger E, Lecommandoux S, Wong AD, Gillies ER, Pedrono B, et al. Thermosensitive polymer-grafted iron oxide nanoparticles studied by in situ dynamic light backscattering under magnetic hyperthermia. Journal of Physics D: Applied Physics. 2015;**48**(49):494001

[71] Espinosa-Cano E, Palao-Suay R, Aguilar MR, Vázquez B, Román JS. Polymeric nanoparticles for cancer therapy and bioimaging. In: Goncalves G, Tobias G, editors. Nanooncology: Nanomedicine and Nanotoxicology. AG: Springer International Publishing; 2018. pp. 137-172

[72] Liu T, Shi S, Liang C, Shen S, Cheng L, Wang C, et al. Iron oxide decorated MoS_2 nanosheets with double pegylation for chelator-free radiolabeling and multimodal imaging guided photothermal therapy. ACS Nano. 2015;**9**:950-960

[73] Harris TJ, von Maltzahn G, Lord ME, Park J-H, Agrawal A, Min D-H. Protease-triggered unveiling of bioactive nanoparticles. Small. 2008;**4**(9): 1307-1312

[74] Fan CH, Cheng YH, Ting CY, Ho YJ, Hsu PH, Liu HL, et al. Ultrasound/ magnetic targeting with SPIO-DOX-microbubble complex for image-guided drug delivery in brain tumors. Theranostics. 2016;**6**(10):1542-1556

[75] Machala L, Tucek J, Zboril R. Polymorphous transformations of nanometric iron(III) oxide: A review. Chemistry of Materials. 2011;**23**: 3255-3272

[76] Tucek J, Zboril R, Namai A, Ohkoshi S. ε-Fe_2O_3: An advanced nanomaterial exhibiting giant coercive field, millimeter-wave ferromagnetic resonance and magnetoelectric coupling. Chemistry of Materials. 2010;**22**: 6483-6505

[77] Kido O, Higashino Y, Kamitsuji K, Kurumada M, et al. Phase transition temperature of γ-Fe_2O_3 ultrafine particle. Journal of the Physical Society of Japan. 2004;**73**(7):2014-2016

[78] Lopez-Sanchez J, Serrano A, Campo AD, Munoz-Noval A, Salas-Colera E, et al. A combined micro-Raman, X-ray absorption and magnetic study to follow the glycerol-assisted growth of epsilon-iron oxide sol-gel coatings. Journal of Alloys and Compounds. 2022;**892**:162061

[79] Chen C-J, Chiang R-K, Lai H-Y, Lin C-R. Characterization of monodisperse Wüstite nanoparticles following partial oxidation. Journal of Physical Chemistry C. 2010;**114**(10): 4258-4263

[80] Ullrich A, Rolle N, Horn S. From wustite to hematite: thermal transformation of differently sized iron oxide nanoparticles in air. Journal of Nanoparticle Research. 2019;**21**:168

[81] Vernekar D, Jagadeesan D. Tunable acid–base bifunctional catalytic activity of FeOOH in an orthogonal tandem reaction. Catalysis Science & Technology. 2015;**5**:4029-4038

[82] Heuss-Aßbichler S, Amthauer G, John M, editors. Highlights in Applied Mineralogy. Berlin: Walter de Gruyter; 2018. ISBN: 978-3-11-049122-7

[83] Tadic M, Spasojevic V, Kusigerski V, Markovic D, Remskar M. Formation of ε-Fe_2O_3 phase by the heat treatment of α-Fe_2O_3/SiO_2 nanocomposite. Scripta Materialia. 2008;**58**:703-706

[84] Zboril R. Iron (III) oxides from thermal processes—Synthesis, structural and magnetic properties, Mossbauer spectroscopy characterization, and applications. Chemistry of Materials. 2002;**14**:969-982

[85] Lopez-Sanchez J, McIntosh G, Osete ML, Campo AD, Villalaın JJ, Perez L, et al. Epsilon iron oxide: Origin of the high coercivity stable low Curie temperature magnetic phase found in heated archeological materials. Geochemistry, Geophysics, Geosystems. 2017;**18**:2646-2656. DOI: 10.1002/2017GC006929

[86] Gracia-Munoz JL, Romaguera A, Fauth F, Nogues J, Gich M. Unveilling a new high-temperature ordered magnetic phase in ε-Fe_2O_3. Chemistry of Materials. 2017;**29**(22):9705-9713

[87] Koch CJW, Borggard OK, Madsen MB, Morup S. Magnetic properties of synthetic Feroxyhite (δ-FeOOH). In: Schultz LG, van Olphen H, Mumpton FA, editors. Proceedings of the International Clay Conference Denver. Virginia, United States: Clay Minerals Society; 1985

[88] Nishida N, Amagasa S, Kobayshi Y, Yamada Y. Synthesis of superparamagnetic δ-FeOOH nanoparticles by a chemical methods. Applied Surface Science. 2016;**387**:996-1001

[89] Akharame MO, Fagbayigbo BO, Pereao O, Oputu OU, Olorunfemi DI, Fatoki OS, et al. Beta-FeOOH nanoparticles: a promising nano-based material for water treatment and remediation. Journal of Nanoparticle Research. 2021;**23**:8

[90] Lee GH, Kim SH, Choi BJ, Huh SH, Chang Y, Kim B, et al. Magnetic properties of needle-like {alpha}-FeOOH and {gamma}-FeOOH nanoparticles. Journal of the Korean Physical Society. 2004;**45**(4):1019-1024

[91] Hiemstra T. Surface structure controlling nanoparticle behavior: magnetism of ferrihydrite, magnetite, and maghemite. Environmental Science: Nano. 2018;**5**(3):752-764

[92] Xia T, Wang J, Wu C, et al. Novel complex-coprecipitation route to form high quality triethanolamine-coated Fe_3O_4 nanocrystals: Their high saturation magnetizations and excellent water treatment properties. CrystEngComm. 2012;**14**(18):5741-5744

[93] Yang HM, Park CW, et al. HER2/neu antibody conjugated poly(amino acid)-coated iron oxide nanoparticles for breast cancer MR imaging. Biomacromolecules. 2010;**11**(11):2866-2872

[94] Lee JH, Huh YM, et al. Artificially engineered magnetic nanoparticles for ultra-sensitive molecular imaging. Nature Medicine. 2007;**131**:95-99

[95] Zhao S, Yu X, Qian Y, Chen W, Shen J. Multifunctional magnetic iron oxide nanoparticles: an advanced platform for cancer theranostics. Theranostics. 2020;**10**(14):6278-6309

[96] Weng Q, Hu X, Zheng J, Xia F, Wang N, Liao H, et al. Toxicological risk assessments of iron oxide nanocluster- and gadolinium-based T_1 MRI contrast agents in renal failure rats. ACS Nano. 2019;**13**:6801-6812

[97] Ajinkya N, Yu X, Kaithal P, Luo H, Somani P, Ramakrishna S. Magnetic iron

oxide nanoparticle (IONP) synthesis to applications: Present and future. Materials. 2020;**13**:4644

[98] Sun C, Veiseh O, Gunn J, Fang C, et al. In vivo MRI detection of gliomas by chlorotoxin-conjugated superparamagnetic nanoprobes. Small. 2008;**4**(3):372-379

[99] Tefft BJ, Uthamaraj S, Harburn JJ, Klabusay M, Dragomir-Daescu D, Sandhu GS. Cell labeling and targeting with superparamagnetic iron oxide nanoparticles. Journal of Visualized Experiments. 2015;**105**:e53099. DOI: 10.3791/53099

[100] Chung S, Revia RA, Zhang M. Iron oxide nanoparticles for immune cell labeling and cancer immunotherapy. Nanoscale Horiz. 2021;**6**:696-717

[101] Shen Z, Chen T, Ma X, Ren W, Zhou Z, Zhu G, et al. Multifunctional theranostic nanoparticles based on exceedingly small magnetic iron oxide nanoparticles for T_1-weighted magnetic resonance imaging and chemotherapy. ACS Nano. 2017;**11**:10992-11004

[102] Arbab AS, Yocum GT, Rad AM, Khakoo AY, Fellowes V, Read EJ, et al. Labeling of cells with ferumoxides-protamine sulfate complexes does not inhibit function or differentiation capacity of hematopoietic or mesenchymal stem cells. NMR in Biomedicine. 2005;**18**(8):553-559. DOI: 10.1002/nbm.991

[103] Korchinski DJ, Taha M, Yang R, Nathoo N, Dunn JF. Iron oxide as an MRI contrast agent for cell tracking. Magnetic Resonance Insights. 2015;**8**(Suppl 1):15-29

[104] Chen TJ, Cheng TH, Chen CY. Targeted herceptin–dextran iron oxide nanoparticles for noninvasive imaging of HER2/neu receptors using MRI. Journal of Biological Inorganic Chemistry. 2009;**14**(2):253-260

[105] Chen TJ, Cheng TH, Hung YC, Lin KT, Liu GC, Wang YM. Targeted folic acid-PEG nanoparticles for noninvasive imaging of folate receptor by MRI. Journal of Biomedical Materials Research. Part A. 2008;**87**(1):165-175

[106] Dadfar SM, Roemhild K, Drude NI, von Stillfried S, Knüchel R, Kiessling F, et al. Iron oxide nanoparticles: Diagnostic, therapeutic and theranostic applications. Advanced Drug Delivery Reviews. 2019;**138**:302-325

[107] Mi Y, Li K, Liu Y, Pu K-Y, Liu B, Feng S-S. Herceptin functionalized polyhedral oligomeric silsesquioxane – conjugated oligomers – silica/iron oxide nanoparticles for tumor cell sorting and detection. Biomaterials. 2011;**32**(32): 8226-8233

[108] Du X, Zhou J, Wu L, Sun S, Xu B. Enzymatic transformation of phosphate decorated magnetic nanoparticles for selectively sorting and inhibiting cancer cells. Bioconjugate Chemistry. 2014; **25**(12):2129-2133

[109] Cheng HW, Tsao HY, Chiang CS, Chen SY. Advances in magnetic nanoparticle-mediated cancer immune-theranostics. Advanced Healthcare Materials. 2021;**10**(1):e2001451. DOI: 10.1002/adhm.202001451

[110] Kircher MF, Allport JR, Graves EE, Love V, Josephson L, Lichtman AH, et al. In vivo high resolution three-dimensional imaging of antigen-specific cytotoxic T-lymphocyte trafficking to tumors. Cancer Research. 2003;**63**(20): 6838-6846

[111] Liu L, Ye Q, Wu Y, Hsieh WY, Chen CL, Shen HH, et al. Tracking T-cells in vivo with a new nano-sized MRI

contrast agent. Nanomedicine. 2012; **8**(8):1345-1354

[112] Li K, Nejadnik H, Daldrup-Link HE. Next-generation superparamagnetic iron oxide nanoparticles for cancer theranostics. Drug Discovery Today. 2017;**22**(9):1421-1429

[113] Li K, Lu L, Xue C, Liu J, He Y, Zhou J, et al. Polarization of tumor-associated macrophages phenotype via porous hollow iron nanoparticles for tumor immunotherapy in vivo. Nanoscale. 2020; **12**:130-144. DOI: 10.1039/c9nr06505a

[114] Li CX, Zhang Y, Dong X, Zhang L, Liu MD, Li B, et al. Artificially reprogrammed macrophages as tumor-tropic immunosuppression-resistant biologics to realize therapeutics production and immune activation. Advanced Materials. 2019;**31**(15):e1807211

[115] de Vries IJ, Lesterhuis WJ, Barentsz JO, Verdijk P, van Krieken JH, Boerman OC, et al. Magnetic resonance tracking of dendritic cells in melanoma patients for monitoring of cellular therapy. Nature Biotechnology. 2005; **23**(11):1407-1413

[116] Zhao Y, Zhao X, Cheng Y, Guo X, Yuan W. Iron oxide nanoparticles-based vaccine delivery for cancer treatment. Molecular Pharmaceutics. 2018;**15**(5): 1791-1799

[117] Traini G, Ruiz-de-Angulo A, Blanco-Canosa JB, Zamacola Bascarán K, Molinaro A, Silipo A, et al. Cancer immunotherapy of TLR4 agonist-antigen constructs enhanced with pathogen-mimicking magnetite nanoparticles and checkpoint blockade of PD-L1. Small. 2019;**15**(4):e1803993

[118] Zhang Q, Wei W, Wang P, Zuo L, Li F, Xu J, et al. Biomimetic magnetosomes as versatile artificial antigen-presenting cells to potentiate T-cell-based anticancer therapy. ACS Nano. 2017;**11**(11):10724-10732

[119] Wang Z, Zhang F, Shao D, Chang Z, Wang L, Hu H, et al. Janus nanobullets combine photodynamic therapy and magnetic hyperthermia to potentiate synergetic anti-metastatic immunotherapy. Advancement of Science. 2019;**6**(22):1901690

[120] Long CM, van Laarhoven HWM, Bulte JWM, Levitsky HI. Magnetovaccination as a novel method to assess and quantify dendritic cell tumor antigen capture and delivery to lymph nodes. Cancer Research. 2009; **69**(7):3180-3187

[121] Kosmides AK, Sidhom J-W, Fraser A, Bessell CA, Schneck JP. Dual targeting nanoparticle stimulates the immune system to inhibit tumor growth. ACS Nano. 2017;**11**(6):5417-5429

[122] Luo X, Peng X, Hou J, Wu S, Shen J, Wang L. Folic acid-functionalized polyethylenimine superparamagnetic iron oxide nanoparticles as theranostic agents for magnetic resonance imaging and PD-L1 siRNA delivery for gastric cancer. International Journal of Nanomedicine. 2017;**12**:5331-5343

[123] Gu Y, Yoshikiyo M, Namai A, Bonvin D, Martinez A, et al. Magnetic hyperthermia with ε-Fe_2O_3 nanoparticles. RSC Advances. 2020;**10**:28786

[124] Kolosnjaj-Tabi J, Di Corato R, Lartigue L, Marangon I, Guardia P, Silva AKA, et al. Heat-generating iron oxide nanocubes: Subtle “destructurators” of the tumoral microenvironment. ACS Nano. 2014; **8**(5):4268-4283

[125] Salimi M, Sarkar S, Saber R, et al. Magnetic hyperthermia of breast cancer

cells and MRI relaxometry with dendrimer-coated iron-oxide nanoparticles. Cancer Nano. 2018;**9**:7

[126] Hayashi K, Nakamura M, Sakamoto W, Yogo T, Miki H, Ozaki S, et al. Superparamagnetic nanoparticle clusters for cancer theranostics combining magnetic resonance imaging and hyperthermia treatment. Theranostics. 2013;**3**(6):366-376

[127] Lin SY, Huang RY, Liao WC, Chuang CC, Chang CW. Multifunctional PEGylated Albumin/IR780/iron oxide nanocomplexes for cancer photothermal therapy and MR imaging. Nano. 2018; **2**(2):106-116

[128] Kirui DK, Rey DA, Batt CA. Gold hybrid nanoparticles for targeted phototherapy and cancer imaging. Nanotechnology. 2010;**21**(10):105105

[129] Yang R, Fu C, Fang J, Xu X, Wei X, Tang W, et al. Hyaluronan-modified superparamagnetic iron oxide nanoparticles for bimodal breast cancer imaging and photothermal therapy. International Journal of Nanomedicine. 2017;**12**:197-206

[130] Espinosa A, Di Corato R, Kolosnjaj-Tabi J, Flaud P, Pellegrino T, Wilhelm C. Duality of iron oxide nanoparticles in cancer therapy: Amplification of heating efficiency by magnetic hyperthermia and photothermal bimodal treatment. ACS Nano. 2016;**10**(2):2436-2446

[131] Chauhan VP, Stylianopoulos T, Boucher Y, Jain RK. Delivery of molecular and nanoscale medicine to tumors: Transport barriers and strategies. Annual Review of Chemical and Biomolecular Engineering. 2011;**2**: 281-298

[132] Jain RK, Stylianopoulos T. Delivering nanomedicine to solid tumors. Nature Reviews. Clinical Oncology. 2010;**7**(11):653-664

[133] Li X, Wang Z, Ma M, Chen Z, Tang X-L, Wang Z. Self-assembly iron oxide nanoclusters for photothermal-mediated synergistic chemo/chemodynamic therapy. Journal of Immunology Research. 2021; Article ID 9958239

[134] Norouzi M, Yathindranath V, Thliveris JA, et al. Doxorubicin-loaded iron oxide nanoparticles for glioblastoma therapy: A combinational approach for enhanced delivery of nanoparticles. Scientific Reports. 2020;**10**:11292

[135] Hussein-Al-Ali SH, Hussein MZ, Bullo S, Arulselvan P. Chlorambucil-iron oxide nanoparticles as a drug delivery system for leukemia cancer cells. International Journal of Nanomedicine. 2021;**16**:6205-6216

Chapter 5

Magnetite Nanoparticles (Fe_3O_4) for Radio-Frequency and Microwave Applications

Poonam Lathiya and Jing Wang

Abstract

The size and shape dependent tunable electromagnetic (EM) properties of magnetite – Fe_3O_4 nanoparticles makes them an attractive material for various future electronics and biomedical device applications such as tunable attenuators, miniaturized isolators and circulators, RF antennas, EM shielding, and biomedical implants etc. The strategic design of RF devices requires specific dielectric and magnetic properties according to the applications, which in turn depends on the size and shape of the particles. At nanoscale, iron oxide's magnetic and dielectric properties are very different from its bulk properties and can be tuned and enhanced by utilizing different synthesis approaches. In this chapter, we summarize electromagnetic properties of magnetite (Fe_3O_4) nanomaterials such as, complex permeability, complex permittivity, magnetic and dielectric loss tangents, saturation magnetization, temperature dependence, and ferromagnetic resonance; and how these properties can be optimized by varying different synthesis parameters. Finally, Fe_3O_4 nanocomposites will be explored by using different synthesis approaches for implementation of RF and microwave applications and we will conclude the chapter with future recommendations.

Keywords: Fe_3O_4 nanoparticles, morphology, magneto-dielectric properties, RF and microwave region, magnetic loss tangent

1. Introduction

Over the past few decades, nanotechnology has expanded its applications exponentially in all aspects of life ranging from biomedical, chemical, material engineering to integrated electronics [1–5]. In nanotechnology, functional nanoparticles with size ranging from 1 to 100 nm have been widely studied [6]. The unique and specifically tailored structure and size dependent properties of the nanoparticles make them extensively important for research and development for various applications such as environment, healthcare, medical, defense, electronics, and so on [7–9]. Nanoparticles have different properties from their bulk counterparts because as the size of the particle decreases, surface effects (more atoms are exposed at the surface of particle, thus leading to highly sensitive and reactive surfaces) and other atomic effects such as quantum confinement effect in electronic structure comes into play [10, 11]. The key to achieve novel chemical, structural, magnetic, physical and mechanical properties of nanoparticles is the large surface to volume ratio [12].

Recently metal oxide nanoparticles such as iron oxide has garnered considerable attention due to its unique structural, electrical, and magnetic properties which, have numerous applications in areas such as data storage, memory devices, water purification, bioprocessing, drug delivery, hyperthermia, magnetic resonance imaging (MRI), biosensors, electronic devices, aerospace applications, etc. [13–16]. Iron oxide is a compound, which can be found in nature in different phases. The most common ones are hematite (α-Fe_2O_3), magnetite (Fe_3O_4) and maghemite (γ-Fe_2O_3) [17, 18]. All of these forms show promising properties such as biocompatibility and relatively low toxicity, high stability under the presence of magnetic field, superparamagnetic response when the particle sizes are kept below 50 nm, and ease of synthesis process and surface treatment [19]. Among them, the magnetite (Fe_3O_4) has been widely used in most practical applications, due to the coexistence of ferrous and ferric cations in Fe_3O_4, which results in fascinating magnetic and structural properties (e.g. high magnetic moment due to valence d electrons) [20]. The high demand and superb performance of magnetite – Fe_3O_4, is attributed to their microstructure, particle size, more active electronic sites, and high surface area to volume ratio [21–24]. The performance of magnetite (Fe_3O_4) nanoparticle properties strongly depends on the oxide phase, the particle morphology, the shape and size distribution, the internal composition (e.g., impurities, grain boundaries and the surface chemistry). Therefore, for a particular application, synthesis methods and procedures must be tailored and optimized [22]. Synthesis methods dictate the crystalline properties, size, shape, and quality of the magnetite nanoparticles. Hence, it greatly affects the magnetic, dielectric and loss properties. Meanwhile, sizes and shapes are also critical; shape change shows crystal facets, and the atomic arrangements in each facet have reflective effects on its electronic properties. There is a growing demand of novel magnetic materials in electronic industry. Here, we will overview and briefly discuss the synthesis methods while focusing on radio frequency (RF) and microwave electronic applications of magnetite (Fe_3O_4) nanoparticles.

Magnetite (Fe_3O_4) nanoparticles can be synthesized using different methods such as physical (laser ablation arc discharge, combustion, electrodeposition, and pyrolysis), chemical (sol–gel synthesis, microemulsion, hydrothermal, coprecipitation, Polyols, thermal decomposition) and biological methods (Protein mediated, plant mediated, bacteria mediated, fungi mediated). Different shapes and sizes of Fe_3O_4 (nanorod, porous nanospheres, nanocubes, distorted cubes, core shell and self-oriented flowers) can be synthesized using same synthesis procedures, by using the optimum synthesis parameters like particular precursor of iron salts, pH levels, and temperature variations etc. [25, 26]. These synthesis methods are easy to implement while playing a major role in controlling the morphology and electromagnetic properties of Fe_3O_4 nanoparticles. In order to make Fe_3O_4 nanoparticles compatible with different applications, proper functionalization and surface modification of Fe_3O_4 is very important [27, 28]. Surface modification of the Fe_3O_4 nanoparticles using different stabilizing agents (PVP, oleic acid, sodium oleate etc.) is a necessary step after or during the synthesis process to make them both biocompatible and stable [29, 30].

For RF and microwave electronics, tunable or reconfigurable devices are becoming important to cause a growing interest of enabling nanotechnology in new wireless devices [31]. Magnetic materials have been used effectively for tunable and reconfigurable of components such as inductors, antennas, and phase shifters [32, 33]. By using tunable properties of Fe_3O_4 nanoparticles in these devices, one can control not only their frequency response but also helpful in improvement of electromagnetic behavior of these devices at a particular frequency [34, 35]. In this chapter, we will discuss the synthesis procedures of magnetite (Fe_3O_4) nanoparticles and their usage in RF and microwave applications. The development of sustainable synthesis approaches for these nanoparticles and investigations of how the structural properties including shape and size of magnetite nanoparticles can enable the tuning of electromagnetic properties for different device applications will be presented.

2. Synthesis methods

As mentioned above, there are different approaches to synthesize magnetite (Fe_3O_4) nanoparticles, which includes physical, chemical, and biological methods. The properties of Fe_3O_4 nanoparticles determine its field of applications. The most widely used synthesis approaches are chemical co-precipitation, thermal decomposition, hydrothermal method, Polyols method and microemulsion method [25].

As shown in the figure, chemical methods are mostly widely used as they are cost effective and easy to handle. Some of the most common synthesis methods are summarized below [25].

2.1 Co-precipitation methods

Co-precipitation synthesis is the most common technique for the synthesis of magnetic magnetite (Fe_3O_4) nanoparticles because of its low cost, environment friendly precursors and simple experimental procedure that occurs at moderately low temperature (20°C - 90°C) [6]. This method is popular because of water based precursor solutions, where simultaneous precipitation of ferrous and ferric ions can occur due to the addition of base in the solution while sustaining a constant pH level. Fe (II) and Fe (III) salts are used in different basic aqueous solutions such as NaOH and NH_4OH to form magnetite (Fe_3O_4) nanoparticles. Nanoparticle size between 5 nm and 20 nm range can be synthesized using this method [11]. Experimental conditions such as Fe^{2+} and Fe^{3+} salt chlorides, sulphates, nitrates, ratio of Fe^{2+} and Fe^{3+} ions in the solution, ionic strength of the solution, pH value of the solution and reaction temperature are very critical parameters to achieve desired size, shape, microstructure, and magnetic properties. Key literature findings about the effects of some of these conditions on nanoparticles properties with a special focus on electronic properties will be detailed below. **Figure 1** shows the typical co-precipitation technique experimental set-up using multistage flow reactor for continuous synthesis of Fe_3O_4 nanoparticles [36].

It is known that co-precipitation method typically results in low saturation magnetization and broad particle size range due to variation in magnetite (Fe_3O_4) nanoparticles core size and agglomeration, which are the main drawbacks [37, 38]. In order to reduce agglomeration and oxidation of Fe_3O_4 nanoparticles, different surface acting reagents and functional materials such as polyethylene glycol (PEG), Polyvinyl Alcohol (PVA), dextrin, Polyvinylpyrrolidone (PVP) etc. can be added during the reaction [39–42].

Radon *et al.* in 2017 studied effect of different organic modifiers (glycol, PVP, citrate and dextrin) on the structural and optical properties of Fe_3O_4 nanoparticles [43]. It was observed that organic modifiers with an exception of glycol facilitate

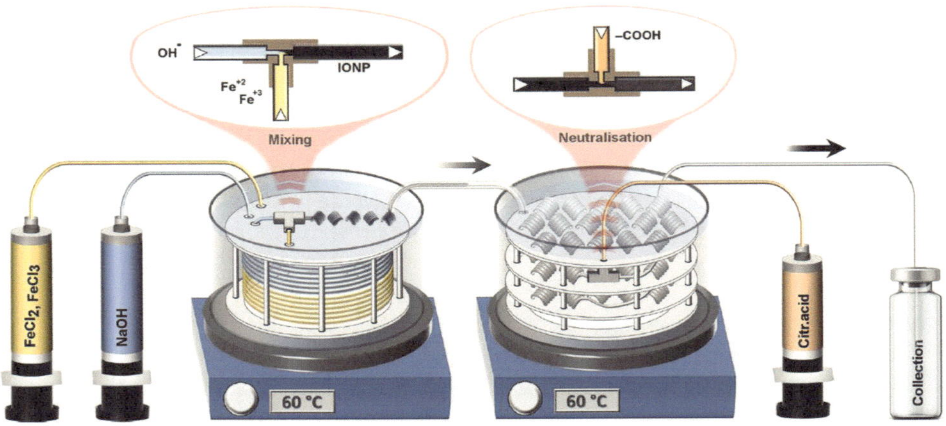

Figure 1.
Co-precipitation method for the synthesis of Fe_3O_4 nanoparticles using multistage flow reactor [36].

reduction of particle size. Particle size range of 2.9 nm to 12.9 nm was reported by modifying co-precipitation process. Fe_3O_4 nanoparticles synthesized with tartaric acid in the solution have the smallest mean particle size along with largest band gap energy [43]. The results showed that an increase in particle size leads to wider optical bandgaps. **Figure 2** shows schematic representation of synthesized Fe_3O_4 nanoparticles along with organic modifiers using XRD, TEM and FTIR data [43].

Similarly, Anbarasu *et al.* studied the effect of PEG on the crystallite size of Fe_3O_4 nanoparticles and saturation magnetization. With an increase in weight of PEG coating (1 g - 3 g) on Fe_3O_4 nanoparticles, crystallite size decreases from 14.96 nm to 10.85 nm, while saturation magnetization also decreases from 62 emu/g to 51 emu/g, accordingly [44].

Saragi *et al.* in 2018 studied the effect of reaction temperature variation on the structure and size of the Fe_3O_4 nanoparticles. They reported an increase in mean particle size from 10.14 nm to 11.66 nm with an increase in reaction temperature from 25 to 80°C [45]. It was reported that smaller crystallite size was measured for low temperature synthesis. It was observed that band gap energy for Fe_3O_4 nanoparticles has decreased from 1.76 eV to 1.14 eV with an increase in particle size. **Figure 3** shows the variation of complex permittivity spectra and magnetization of Fe_3O_4 nanoparticles with respect to temperature variation [46]. There was a nonlinear relation between complex permittivity and temperature. Similar effects of temperature variation on dielectric and structural properties of Fe_3O_4 nanoparticles have been reported by Radon *et al.* in 2018 [46]. The permittivity and dielectric loss as seen in **Figure 3** exhibited a nonlinear behavior in response to temperature variation [45]. It is observed the dielectric property was mostly dominated by polarization process. The electromagnetic shielding mechanism of nanocomposites can be attributed to absorption process of low-reflection electromagnetic shielding composites.

Optimization of co-precipitation synthesis parameters in order to control the particle size and polydispersity can be quite challenging, extensive ongoing research have been carried out to understand the mechanism of particle formation so that particle structures/properties can be tailored for applications.

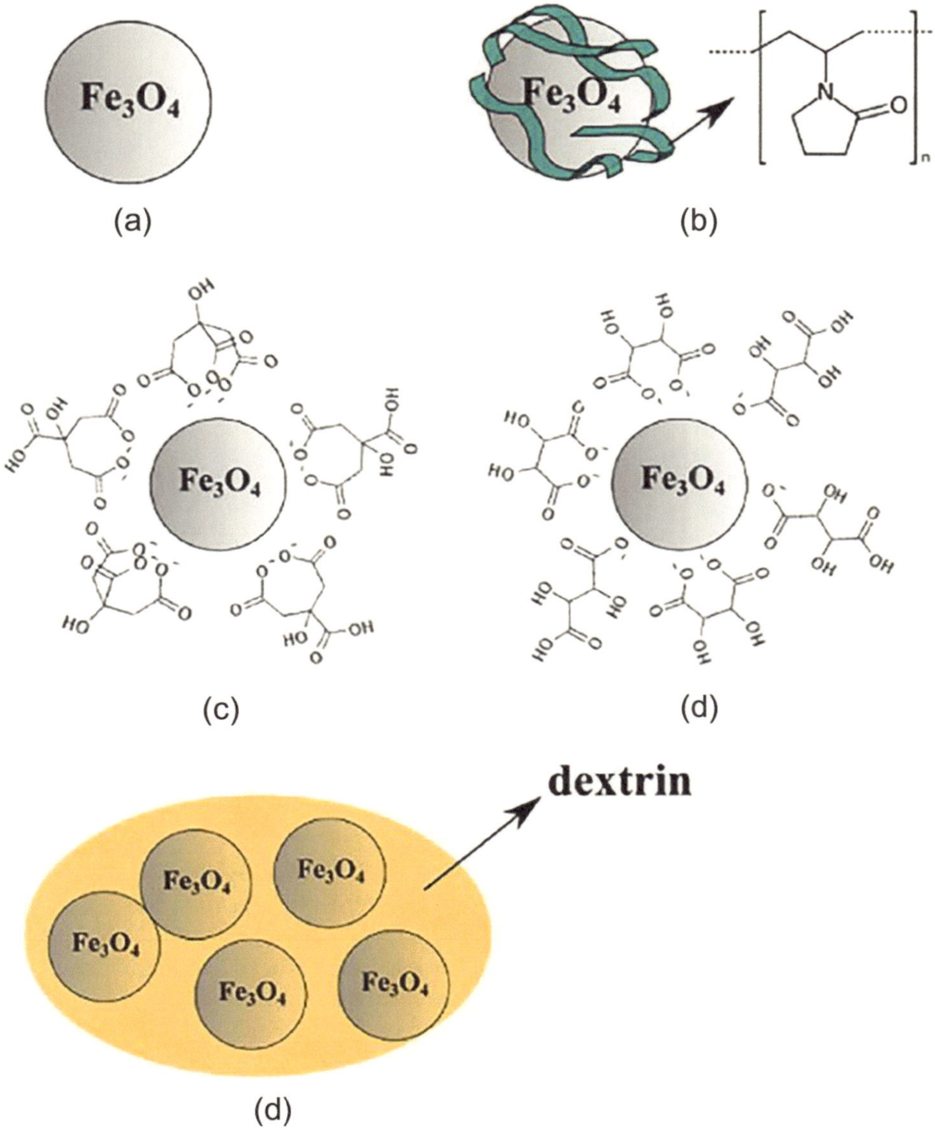

Figure 2.
A schematic representation of synthesized Fe_3O_4 nanoparticles along with organic modifiers using XRD, TEM and FTIR data by showing (a) Fe_3O_4 and glycol; (b) Fe_3O_4 and PEG; (c) Fe_3O_4 and citrate; (d) Fe_3O_4 and tartrate; and (e) Fe_3O_4 and dextrin [43].

2.2 Thermal decomposition

Thermal decomposition is a synthesis of Fe_3O_4 nanoparticles using decomposition of iron precursor at high temperature in organic phase solution [47]. In this method, precursors of iron (III) acetylacetonate, $Fe(acac)_3$, iron nitro sophenylhydroxylamine or iron pentacarbonyl are used in oleic acid or lauric acid, which are oxidized at high

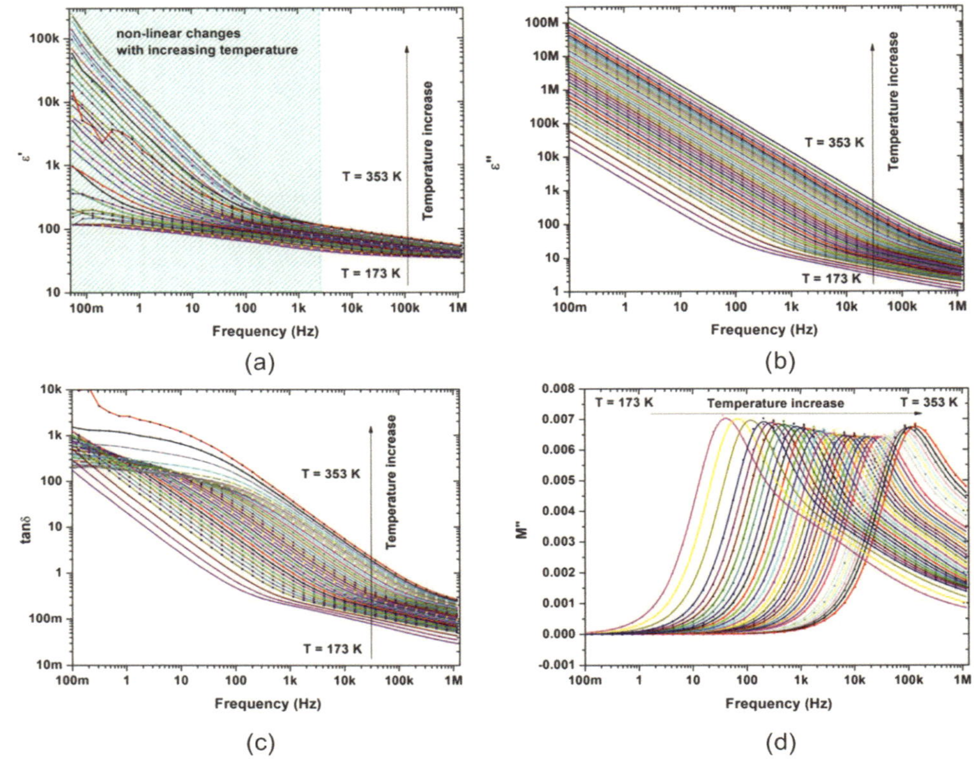

Figure 3.
Variation of complex permittivity spectra and magnetization of Fe_3O_4 nanoparticles with respect to temperature variation [46].

temperature to make monodisperse Fe_3O_4 nanoparticles [6]. **Figure 4** presents a conceptual illustration of experimental process to synthesize of monodisperse Fe_3O_4 nanoparticles [47].

The thermal decomposition method can be used to synthesize monodisperse nanoparticles of up to 20 nm in size with a tight size distribution. Wetterskog *et al.* in 2015 have synthesized nanospheres and nanocubes shaped Fe_3O_4 nanoparticles using thermal decomposition method [48]. A size distribution between 5 nm and 27 nm was reported for both nanocubes and nanospheres. Their work demonstrated that size of nanoparticles can be tuned by reaction temperature and shape can also be tuned by addition of oleic acid or sodium oleate during the synthesis. **Figure 5** shows variation of shape from nanocube to nanosphere with the addition of sodium oleate [48]. Similar effects of oleic acid or sodium oleate on the shape of Fe_3O_4 nanoparticles have been reported in the literature [49–52]. Li *et al.* in 2010 fabricated oleic acid coated Fe_3O_4 nanoparticles heated at 320°C under nitrogen atmosphere with uniform shapes and sizes [53]. To address the stabilization issue, different stabilizing agents such as bilayer oleic acid, fatty acids, phenol etc. were studied in the past [47, 54, 55]. For example, Wang *et al.* in 2012 used phenol as a reducing agent and stabilizer for the formation of stable water soluble Fe_3O_4 nanoparticles [56]. The main disadvantage of this method is it requires nanoparticles to dissolve in nonpolar solution for storage and cannot be scaled up for industrial production [57]. Using thermal decomposition methods, high saturation magnetization values with low coercive fields can be achieved by using high

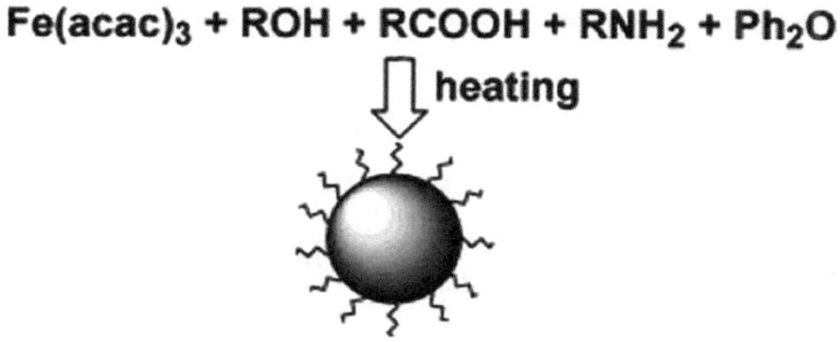

Figure 4.
Conceptual illustration of synthesis process of monodisperse Fe_3O_4 nanoparticles using thermal decomposition method [47].

Figure 5.
Variation of shape from nanocube to nanosphere vs. adjusted addition of sodium oleate [48].

reaction times in the process. *Vuong* et al. in 2015 showed high values of saturation magnetization up to 70 emu/g with reaction time of 120 mins [58].

2.3 Polyol method

Polyol method is a well-known technique to synthesize defined shape and size-controlled metallic, oxide, and semiconductor nanoparticles such as magnetite (Fe_3O_4) nanoparticles [25]. This method involves chemical reduction of metal salts in polyols such as polyethylene glycol at high temperature. The average size of these nanoparticles can be controlled by reactive mediums and this method is widely used to obtained nanoparticles Of size up to 100 nm [21]. The shape, size, particle growth and yield depend upon the type of polyols, salt ratio, concentration, and other physiological condition. Polyol and polyethylene glycol are normally used as solvents, which can dissolve inorganic compounds and offer a wide range of temperature for the reaction. Polyols act as both stabilizer and reducing agent in the reaction and help in prevention of agglomeration and control of particle growth [59]. Abbas *et al*. in 2013 studied the effect of polyethylene glycol as the stabilizer and reducing agent for the synthesis of hydrophilic, monodisperse superparamagnetic Fe_3O_4 nanoparticles for biomedical applications [60]. Similar prior studies have been reported using different polyols such as ethylene glycol, di ethylene glycol, tri ethylene glycol, tetra ethylene glycol and propylene glycol, polyethylene glycol [61–65].

There are also a variety of prior works in the literature, which utilizes solvothermal polyols method to synthesize different Fe_3O_4 cluster sizes for better magnetic properties such as saturation, magnetization and coercivity. In solvothermal polyol method, Fe_3O_4 clusters can be prepared by change of reaction conditions of the solvothermal process and by utilizing sodium acetate [66]. Leung *et al.* in 2009 synthesize Fe_3O_4 clusters of different size by varying the reaction conditions [67]. They are the first group that studied the effect of solvent composition on Fe_3O_4 cluster size. Similar studies have been reported by different groups on controlling the size of Fe_3O_4 clusters [68], core/shell (Fe_3O_4/ZnO) submicron particles [69], PAA modified hydrophilic Fe_3O_4 nanoparticles [70] and Fe_3O_4 nanoparticles [71].

Sayed *et al.* in 2015 utilized microwave assisted solvothermal polyols to synthesize six different shaped Fe_3O_4 nanoparticles using different iron salt as precursors – nanorods, nanohusk, distorted nanocubes, nanocubes, porous spheres and self-oriented flowers [72]. These shapes of Fe_3O_4 were synthesized using KCC-1 synthesis protocol [73], which has involved various iron salts as precursors, cetyltrimethylammonium bromide (CTAB) as a template, along with utilization of cyclohexane-water-pentanol as a reaction solvent and urea as hydrolyzing agent. **Figure 6** shows the SEM images of different shaped of Fe_3O_4 nanoparticles [72].

2.4 Hydrothermal method

Hydrothermal synthesis is the most commonly used method for the preparation of nanomaterials. This is a solution reaction-based approach, which utilizes a wide temperature range from room temperature to high temperatures [74]. To control the morphology of the nanoparticles, low-pressure or high-pressure conditions can be used in the reaction. Pressures above 2000 psi needs to be maintained in hydrothermal synthesis method [25]. The compositions, morphology, particle size of nanomaterials to be synthesized can be well controlled by temperature variation in combination with right precursors in hydrothermal synthesis through liquid phase or multiphase chemical reactions. The particle size and size distribution can also be controlled with precursor concentration [21]. The main drawback of this method is that it needs expensive reactors [1].

Figure 6.
SEM images of six different shaped of Fe_3O_4 nanoparticles obtained by microwave assisted solvothermal polyol method by using KCC-1 synthesis protocol, including: (a) Nanorod, (b) Nanohusk, (c) distorted cubes, (d) Nanocubes, (e) porous spheres, and (f) self-oriented flowers [72].

Gomez *et al.* in 2019 studied the effect of temperature on the morphology of the hydrothermally synthesized Fe_3O_4 nanoparticles. The shape was controlled by the temperature of the reaction; at 120°C, 140°C, and 160°C, to obtain quasi-spheres, octahedrons, and cubes, respectively [75]. **Figure 7** shows the SEM images of Fe_3O_4 nanoparticles obtained at different temperatures [75]. Similar studies were focused on controlling the shape and size of the Fe_3O_4 nanoparticles by controlling temperature, solvent, precursor salt, reducing agent, and so on, while using this hydrothermal method [76–80].

2.5 Microemulsion method

Microemulsion is an isotropic and thermodynamically stable single phase formed by mixing oil, water and surfactants; where oil and water are immiscible, and surfactant has an amphiphilic behavior [81]. There are three main categories of microemulsions - oil in water, water in oil and bi-continuous [1]. Microemulsion method has been known

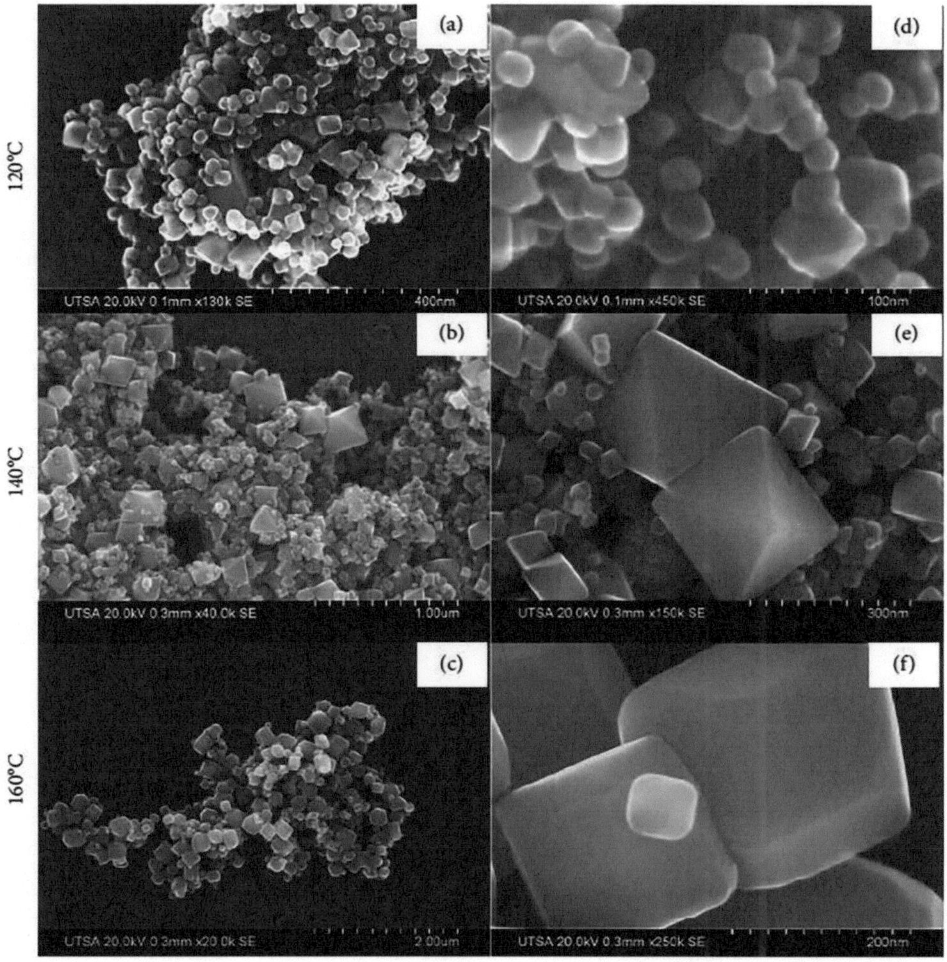

Figure 7.
SEM images of Fe_3O_4 nanoparticles synthesized using hydrothermal method at - (a) 120°C, (b) 140°C, (c) 160°C, where (d) to (f) are zoomed-in SEM photos of the nanoparticles at the corresponding temperatures [75].

to produce narrow particle size distribution between 4 and 15 nm with different shapes. Synthesis of Fe_3O_4 nanoparticles with controlled size and shape can be carried out in water-oil microemulsion, which consists of cationic or non-ionic surfactant (Triton-X), a co-surfactant (n-hexanol, glycols, 1-butanol), oil phase (n-heptane, n-octane, cyclohexane) and aqueous phase. Microemulsion can be carried out through addition of aqueous solution with iron precursor to the surfactant mixture [6]. The major drawback of this method is that the scale up of this method from laboratory scale to mass production at industrial levels could be difficult; particle size and shape changes significantly at large scale despite maintaining the same reaction conditions as lab experiments.

Many prior studies have been reported on the controlled synthesis of Fe_3O_4 nanoparticles using microemulsion method [82–85]. In order to increase the stability of Fe_3O_4 nanoparticles and avoid agglomeration, they have been encapsulated

Methods	Size (nm)	Shape	Saturation Magnetization Ms. (emu/g)	Advantages	Disadvantages
Co-precipitation	3–100	Spherical	20–80	Low to mild temperature, high yield, scalable, inexpensive synthesis, simple purification	Agglomeration, polydispersity
Thermal decomposition	3–80	Spherical, 1D and 2D	Less than 90	Narrow size distribution, high crystallinity, size and shape control	Long reaction time, high temperature, organic medium, expensive, low yield
Polyols Method	10–1000	0D,1D,2D,3D	20–120	size and shape control, less agglomeration, high yield,	Broad particle size distribution
Hydrothermal	2–1000	0D,1D,2D,3D	20–110	High purity nanoparticles, medium temperature, low cost, use stabilizers in reaction to control agglomeration, high yield, aqueous reaction medium	Long reaction time, broad particle size distribution
Microemulsion	4–50	Spherical and cubic	30–110	Low temperature, ambient atmosphere, narrow size distribution, controllable size	Long reaction times, agglomeration, low yield, difficult to remove surfactants

Table 1.
Comparison between different synthesis methods of magnetite (Fe_3O_4) nanoparticles.

with silica precursor, which significantly increase the stability of nanoparticles and protecting them from oxidation [79, 86, 87]. Asab *et al.* in 2020, reported silica coated Fe_3O_4 nanoparticles using water-in-oil microemulsion method for the application of antimicrobial activity. It was observed that silica-coated Fe_3O_4 nanoparticles exhibited homogeneous distribution of particles with relatively less severe agglomerate of the particles [86]. Comparison among different synthesis methods is presented in **Table 1**.

3. Application of magnetite (Fe_3O_4) nanoparticles

Magnetite (Fe_3O_4) nanoparticles are well suited for a wide variety of scientific and engineering applications in numerous fields, due to their strong superparamagnetic and surface properties. Detailed application areas are summarized in **Table 2**. We herein specifically focus on radio frequency (RF) and microwave applications.

Area	Applications
Biomedical and healthcare	Drug delivery [88–90], magnetic hyperthermia [91–94], MRI imaging [42, 95, 96], magnetic separation, controlled drug release, cellular therapy, cell separation and handling of cells [97, 98], purifying cell populations, diseases of the musculoskeletal system, severe inflammation, toxicity [99]
Agriculture	Nano fertilizers, nano fungicides, nano pesticides [100, 101]
Environment	Wastewater treatment, catalyst coatings [102–104]
Recording and storage	Ferrofluids, external magnets [105]
Industries	Catalyst [106, 107]
Textile	Nanofibers, sensors, smart materials [108–110]
Defense	Sensors, nanocomposites, smart materials [111, 112]
Electronics	Printed electronics, spintronics and quantum dots [113, 114]

Table 2.
Scientific and engineering fields of applications for magnetite (Fe_3O_4) nanoparticles.

4. Application of magnetite (Fe_3O_4) nanoparticles for RF and microwave region

With the continuous technological advancements and emerging applications in biomedical devices and electronics in RF and microwave regions, the strategic design of suitable electromagnetic materials requires controlled and well-tailored dielectric, magnetic and loss properties. There is a growing demand to increase the operating frequency of RF and microwave devices. Magnetite (Fe_3O_4) nanoparticles have recently shown great promises for these applications due to their exciting and superior magnetic properties at high operating frequencies [35]. Nevertheless, as an emerging research area with an aim to employ Fe3O4 nanoparticles for unique RF applications, there are relatively limited prior works at this stage.

Fe_3O_4 nanomaterial is the among the very few magnetic materials that exhibits excellent tunable properties using different synthesis approaches. Fe_3O_4 nanoparticles have attracted considerable attentions because of its shape and size tunability, which in turn impact the magnetic and loss properties. The tunable electromagnetic properties of Fe_3O_4 nanoparticles are uniquely suited for designing RF/microwave devices due to their structural and size dependent magnetic and dielectric properties, which can further tuned by external magnetic fields [115, 116]. Meanwhile, self-biased soft magnetic ferrites have been recently explored to exhibit unique properties by exploiting the anisotropy of magnetic material [117–119]. Fe_3O_4 nanoparticles polymer composites have exhibited unique attributes for biomedical device and electronic applications, which require tuned, light weight, robust, flexible and cost-effective devices such as antennas [120].

4.1 Tunable electromagnetic properties of magnetite (Fe_3O_4) nanoparticles for RF and microwave devices

In 2008, Kuanr *et al*. studied the size dependent magnetic properties of Fe_3O_4 nanoparticles for microwave devices [121]. To study the electromagnetic properties of Fe_3O_4 nanoparticles, nanoparticles fluid with oleic acid was spin-coated on a GaAs substrate equipped with a Cu coplanar waveguide (CPW). **Figure 8** shows the transmission response and frequency shift with varied Fe_3O_4 nanoparticle sizes. They reported an increase in resonance frequency with an increase in particle size from 4 nm to 14 nm and then starts to decrease as shown in **Figure 8** (c) [121]. Change in magnetization from Ms = 0.14, 0.19, 0.25, and 0.29 kG for the 4, 6, 8, and 10 nm nanoparticle samples, respectively, was reported. These values are consistent with other reported prior works. The Fe_3O_4 nanoparticles used were synthesized by thermal decomposition method.

Recently, Jadav *et al*. in 2020 studied the effect of particle size on the microwave behavior of Fe_3O_4 nanoparticles in RF region (250 MHz-3 GHz) [122]. It was observed that maximum loss tangent increases, and minimum reflection loss (RL) decreases with a increment of mean particle size. For instance, the maximum magnetic loss was increased by 55.6% and minimum RL was decreased by 34.5% by increasing the mean size of Fe_3O_4 nanoparticles from 11 nm to 16 nm. **Figure 9** shows the magnetic loss tangent and return loss variation with frequency for 4 different nanoparticle sizes (MF1–10 nm to MF4–17 nm) in magnetic fluid [122]. The maximum loss tangent and minimum RL was achieved for 17 nm particle size (MF4) because of high magnetic permeability resulted from a large mean particle size. These properties of Fe_3O_4 nanoparticle are affected by an external magnetic field strength between 0 Oe and 380 Oe, while retaining particle size. For example, resonance frequency shifted form 1.28 GHz (0 Oe) to 2.03 GHz (368 Oe) by applying a 368 Oe magnetic field, accompanied by an increase in magnetic loss tangent for a 10 nm particle sized sample. This was attributed to the fact that shape and size distribution of nanoparticles leads to change in anisotropy constant, which in turn affect the magnetic permeability and magnetic losses. As the magnetic field strength increases, resonance frequency shifts to higher value. Also, the bandwidth increases for largest particle size. The magnetic loss tangent decreases with an increase in magnetic field strength for the largest particle size, while the field strength in the nanoparticle fluid reaches to the maximum and then drop to low value for small size particle. The Fe_3O_4 nanoparticles studied in this work were synthesized using co-precipitation method. Different pH variations were used to synthesize different size nanoparticles at constant temperature for the measurement of these microwave properties [122].

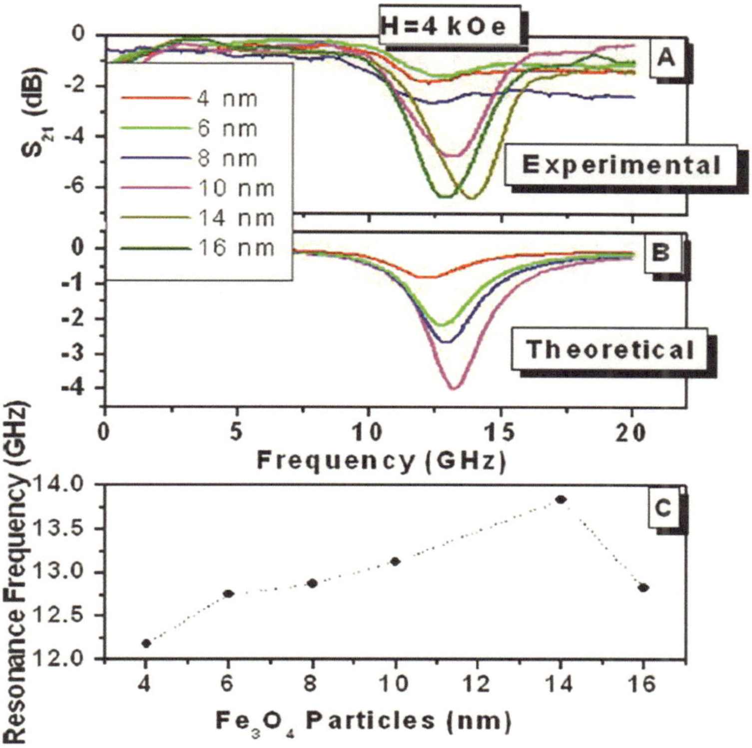

Figure 8.
(a) Measured transmission responses vs. particle sizes under a 4 kOe of external magnetic field; (b) theoretical model-predicted transmission responses vs. particle sizes under a 4 kOe of external magnetic field; and (c) measured resonance frequency vs. Fe_3O_4 particle size [121].

Similarly, the effect of particle concentration, external magnetic field, frequency dependence of RF and microwave properties [35, 123, 124], agglomeration effects on the effective electromagnetic properties of composites with magnetic Fe_3O_4 nanoparticles [125, 126] have been studied and reported in the literature. For example, Li et al. in 2015 reported water soluble Fe_3O_4 nanoparticles coated using surface double-layered self-assembly method. The sodium alpha-olefin sulfonate (AOS) was used as the coating material for better superparamagnetic properties [127]. It was confirmed that AOS double coated Fe_3O_4 magnetic nanoparticles showed less agglomeration as compared to Fe_3O_4 nanoparticles. Saturation magnetization value of about 44.45 emu/g and the blocking temperature T_B 170 K were reported for Fe_3O_4-AOS capped nanoparticles which are ideal values for biomedical applications.

Fabrication of heterostructures is another way to tailor the magnetic properties of the soft magnetic ferrites such as the ones based on Fe_3O_4 nanoparticles for planar device applications (e.g., inductors and patch antennas) including multi-layer ferrite materials with isostructural and non-isostructural materials, (e.g., Fe_3O_4/NiO, Fe_3O_4/CoO, (Mn, Zn)Fe_2O_4/$CoFe_2O_4$, etc.). The combination of Fe_3O_4 soft magnetic ferrite layer and a piezoelectric layer can lead to new and exciting RF and microwave applications such as antenna, sensors etc. [20].

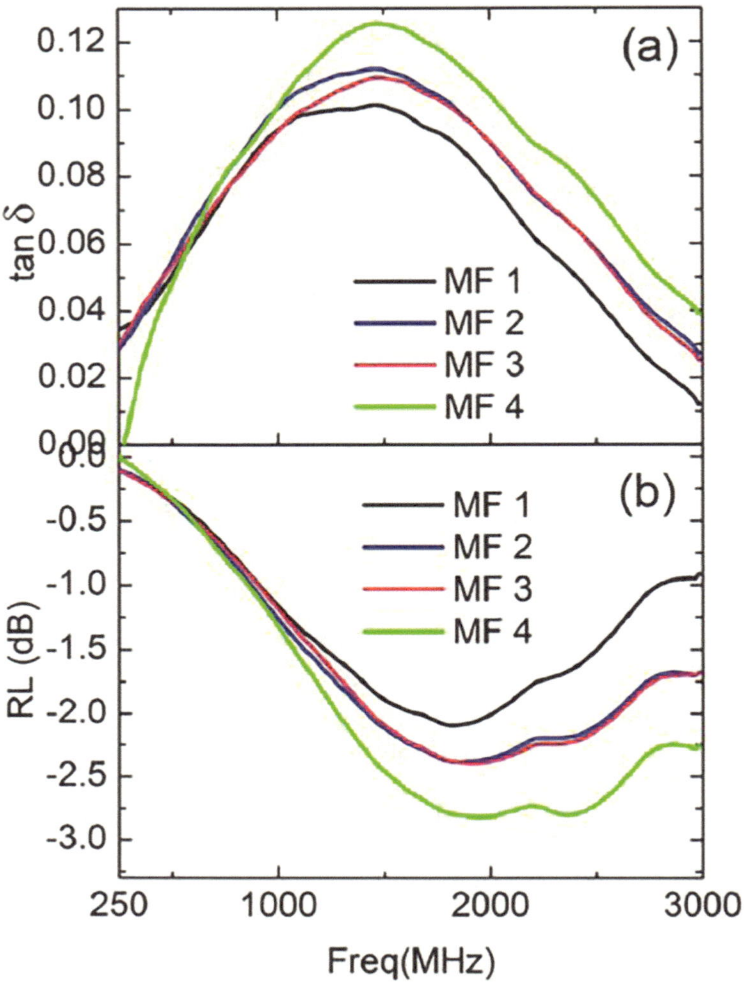

Figure 9.
(a) Variation of magnetic loss tangents vs. frequency for a variety of samples with varied sizes of Fe_3O_4 nanoparticles; (b) return loss variation vs. frequency for 4 samples with different nanoparticle sizes in magnetic fluid. MF1, MF2, MF3, MF4 are magnetic fluids with 10 nm, 12 nm, 16 nm, and 17 nm Fe_3O_4 nanoparticles, respectively [122].

4.2 Magnetite (Fe_3O_4) nanoparticles composite for microwave absorption applications

With rapid advancements in science and technology, the use of RF and microwave electronics have increased many folds, which creates electromagnetic interference (EMI) to not only impact human health but also interfere with electronics nearby [128]. Thus, electromagnetic (EM) absorption materials at RF and microwave frequencies have garnered a great deal of attentions because of their application in wireless data communication, radar system and other area networks [126]. For good microwave absorption properties, impedance matching between air and absorbing material as well as reflection loss are very important. Materials that have both desired magnetic and dielectric properties serve this purpose well [129]. Currently, soft magnetic ferrites and nanomaterials have widely explored for microwave absorption because of their high magnetic, electric and loss properties [130, 131]. Fe_3O_4 is well known for its chemical stability and tailorable magnetic/dielectric losses at microwave regions. Developing low-density composites of high dielectric and magnetic losses as absorbing materials is an effective approach for fulfilling EM absorption performance.

In 2007, Zhou *et al.* reported microwave absorption properties of $SiC@SiO_2@Fe_3O_4$ hybrids for the frequency range of 2–18 GHz for samples of different thickness [132]. $SiC@SiO_2$ nanowires were synthesized using carbothermal reduction method. $SiC@SiO_2@Fe_3O_4$ hybrids were produced by adding iron precursor, $Fe(acac)_3$ to the $SiC@SiO_2$ suspension in triethylene glycol, which is then heated to 280°C under an argon atmosphere. Microwave absorption properties were investigated for different $SiC@SiO_2$ to $Fe(acac)_3$ mass ratios. The EM absorption performance was enhanced by attaching Fe_3O_4 nanoparticles to $SiC@SiO_2$ nanowires. Specially for 1:3 mass ratio of SiC@SiO2 to iron (III) acetylacetonate, the measured microwave absorption for a 2-mm thick sample exhibited a minimum reflection loss of −39.58 dB at 12.24 GHz. **Figure 10** shows reflection loss of $SiC@SiO_2$ and $SiC@SiO_2@Fe_3O_4$ hybrids with different mass ratios [132].

In 2007, Qiao *et al.* studied nanochains yolk-shell Fe_3O_4@N-doped carbon as a novel microwave absorption material for the frequency range of 2–18 GHz [128]. First, the core-shell Fe_3O_4@P(EGDMA-MAA) were synthesized using magnetic field induced precipitation polymerization method, then core-double-shell Fe_3O_4@P(EGDMA-MAA)@PPy nanochains were synthesized using oxidant-directed vapor-phase polymerization process. Finally, yolk-shell Fe_3O_4@N-doped carbon nanochains were synthesized by carbonized Fe_3O_4@P(EGDMA-MAA)@PPy nanochains utilizing salt crystallization method. It was reported that 20% loading of yolk-shell Fe_3O_4@N-doped carbon with paraffin-based composites have exhibited strong

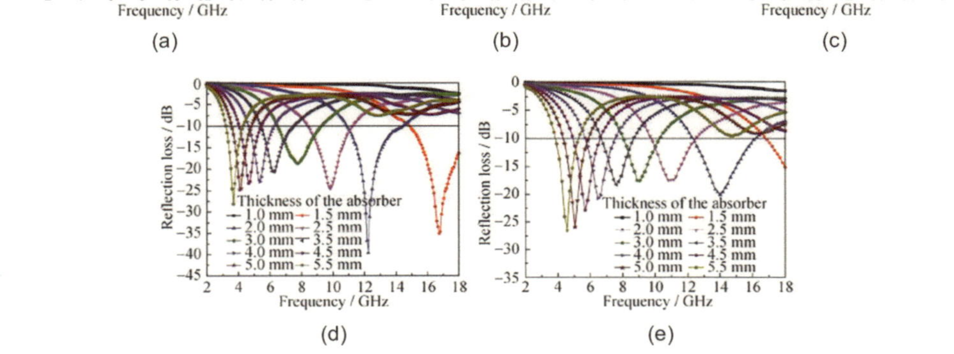

Figure 10.
Calculated reflection loss of (a) $SiC@SiO_2$ nanowires; (b) $SiC@SiO_2@Fe_3O_4$ hybrids in ratio 1:1; (c) 1:2; (d) 1:3; (e) 1:4 [132].

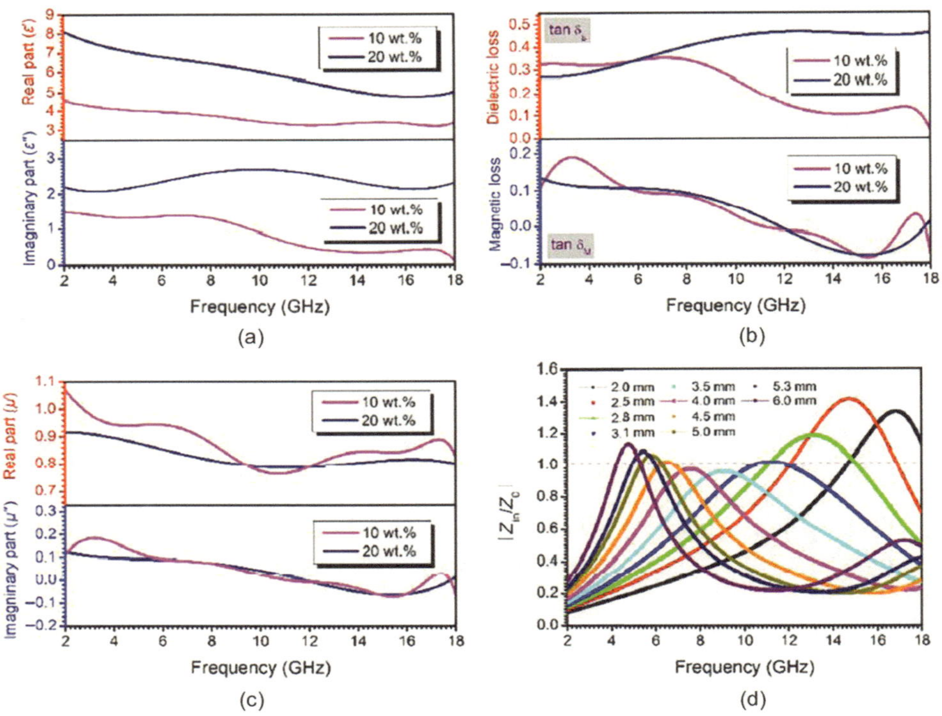

Figure 11.
Electromagnetic parameters of Fe_3O_4@N-doped carbon nanochains including: (a) complex permittivity; (b) dielectric and magnetic loss; (c) complex permeability; and (d) relative input impedance at different thickness layers [128].

absorption capability with a low reflection loss of −63.09 dB at 11.91 GHz. **Figure 11** shows the complex permeability and permittivity for 10 wt % and 20 wt% loading yolk-shell Fe_3O_4@N-doped carbon samples and relative input impedance for samples with different layer thicknesses [128].

The frequency dependent complex relative permeability is given by Eq. (1) [133],

$$\mu_r = \mu' - j\mu'' \tag{1}$$

where μ_r is the ratio of the relative permeability of the material versus that of the free space (μ_0). μ' and μ'' are real and imaginary parts of the relative permeability, respectively.

The magnetic loss tangent is the ratio between the real and imaginary parts given by Eq. (2),

$$\tan\delta_m = \frac{\mu''}{\mu'} \tag{2}$$

The frequency dependent relative complex permittivity can be given by Eq. (3) [133],

$$\varepsilon_r = \varepsilon' - j\varepsilon'' \tag{3}$$

where ε_r is known as ratio of the permittivity of the material versus that of the free space (ε_0) and ε' and ε'' are real and imaginary parts of the complex permittivity.

The dielectric loss tangent is given by Eq. (4),

$$\tan\delta_\varepsilon = \frac{\varepsilon''}{\varepsilon'} \tag{4}$$

The samples with 20 wt% loading showed the highest relative permittivity (real part) along with high dielectric loss tangent over the entire frequency range, which can be ascribed to the conductive loss inside the nanochain during the propagation of electromagnetic wave through the yolk-shell structure. Due to the geometry of yolk-shell structure, such as high porosity and void spaces, multiple scattering and reflections are generated through the interface polarization, which influences the dielectric loss of the nanochains [128]. It was concluded that high magnetic losses (due to natural resonance and eddy current effect) and dielectric losses (due to interfacial polarization) can be achieved by designing porous magnetic cores with proper yolk shell structure. Hence, better microwave absorption performance can be achieved even at low filler loadings.

Similar prior works using Fe_3O_4 nanoparticles as core material have reported recently. **Table 3** tabulated the microwave absorption performance of Fe_3O_4 nanoparticles-based nanocomposites used with different structures.

4.3 Magnetite (Fe_3O_4) nanoparticles composite for antennas in RF and microwave regions

Tunable electromagnetic properties of nanomaterial-based nanocomposite are key enabler for RF and microwave applications. Several reports have described the development of RF and microwave device applications, such as antennas, and inductors using commercially available dielectric and semiconductor-based substrates. For tunable electronic devices, magnetic nanocomposites can facilitate in designing of fully tunable and magnetically controllable devices. This kind of application requires antennas and other RF devices to be operating at different frequencies or meeting other performance needs such as antenna bandwidth and efficiency. RF devices that are frequency agile or dependent are highly desirable for biomedical and defense applications. Tuning of different parameters of device such as frequency can

Absorbers	Absorber Thickness (mm)	RL_{min} vs. (frequency)	Absorption bandwidth (GHz)	References
PANI/ Fe_3O_4	1.4	−18 dB (8.6 GHz)	—	[134]
ACV/Fe_3O_4	2	−30.7 dB (16.4 GHz)	8.2	[131]
RGO/SiO_2/Fe_3O_4	4.5	−56.4 dB (8.1 GHz)	7.1 (3 mm)	[135]
Fe_3O_4/RGO	3.5	−45 dB (8.96 GHz)	3.2	[126]
Fe_3O_4/ZnO	3.5	−22.7 dB (13 GHz)	5.9	[136]
Fe_3O_4/Ppy/CNT	3	−25.9 dB (10.2 GHz)	4.5	[137]
Fe_3O_4/C	2.9	−46 dB (12.8 GHz)	6.5	[138]
Fe_3O_4/TiO_2	2	−23.3 dB (7 GHz)	5.5	[139]

Table 3.
Microwave absorption performance of Fe_3O_4 nanoparticles-based nanocomposites.

be achieved by various methods. One such method for controlling the performance of RF microwave devices is employing tunable magnetic materials such as Fe_3O_4 nanoparticles nanocomposite as the base substrates.

Morales *et al*. in 2011 and 2014 reported implementation of magnetite (Fe_3O_4) nanoparticles and -polydimethylsiloxane (PDMS) magnetodielectric composite for RF and microwave applications [140, 141]. In this work, sub-10 nm Fe_3O_4 nanoparticles have been synthesized using thermal decomposition method. Negligibly low hysteresis losses were reported at room-temperature. Magnetic and dielectric properties of Fe_3O_4-PDMS nanoparticles composite were extracted using multilayer microstrip line test fixtures with and without an external magnetic field of varied levels. Three different concentrations (30, 50 and 80 wt%) of Fe_3O_4 nanoparticles filler in PDMS matrix were studied for both dielectric and magnetic measurements. A relative permeability of 2.5 along with a magnetic loss tangent of 0.15 at 4 GHz was reported for 80 wt% nanocomposite without the application of external magnetic field [141]. Similarly, a permittivity of 2.8 with a dielectric loss tangent of 0.18 were reported for sample with 80 wt% loading at 4 GHz. **Figure 12** shows the room temperature magnetization (M-H) curve for Fe_3O_4 nanoparticles and Fe_3O_4-PDMS nanoparticle composites at three different nanoparticles loading (30, 50 and 80 wt%) [141]. No magnetic hysteresis was observed for all three concentrations of Fe_3O_4-PDMS nanoparticles composite, which is a desirable property for RF microwave antenna applications.

Enhanced permeability and permittivity values of 3.55 and 2.79 along with low magnetic and dielectric loss tangents of 0.02 and 0.019, respectively, were measured for samples with a high loading ratio (80 wt%) of Fe_3O_4 nanoparticles for the composite samples under an external applied field of 0.2Telsa. Based on the optimal magnetic and dielectric properties of nanocomposite under external field polarization, the Fe_3O_4-PDMS nanocomposites have been used to form the substrate for miniaturized multilayer patch antennas with a center frequency of 4GHz, which showed 58% bandwidth enhancement and 57% of size reduction as compared those of PDMS substrate based counterparts. Meanwhile, a return loss of −23 dB and an antenna gain of 2.12 dBi have been achieved. **Figure 13** shows the schematic of multilayer microstrip patch antenna designed with a Fe_3O_4-PDMS composite substrate with a 80 wt% Fe_3O_4 filler loading [140].

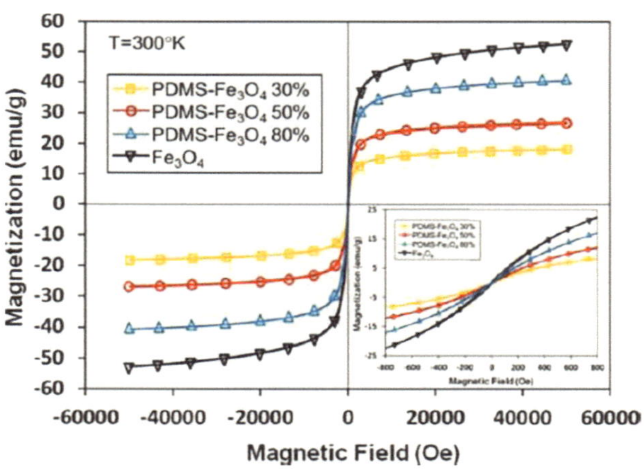

Figure 12.
Magnetization vs. magnetic field (M-H) curve for Fe_3O_4 nanoparticles and Fe_3O_4-PDMS nanoparticles composite at three different nanoparticles loading (30, 50 and 80 wt%) [141].

In 2016, Alqadami *et al.* also reported similar results based on Fe_3O_4-PDMS nanoparticle composite as magneto-dielectric substrate for MIMO antenna array [142]. The reported antenna was designed with 35%- 65% Fe_3O_4 nanoparticles to PDMS polymer ratio over the frequency range of 5.33–7.70 GHz. The reported results showed a bandwidth enhancement of 40.8% and 57% enhancement in radiation efficiency along with an antenna gain of 9.95 dB gain and a return loss of −33 dB. **Figure 14** shows the fabricated prototype of 2x4 MIMO antenna array [142].

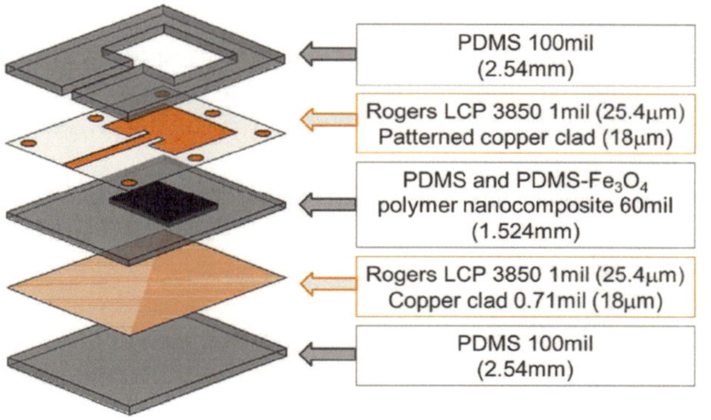

Figure 13.
Real permeability of Fe_3O_4-PDMS nanoparticles composite at varied concentrations of Fe_3O_4 nanoparticles under application of external magnetic field [140].

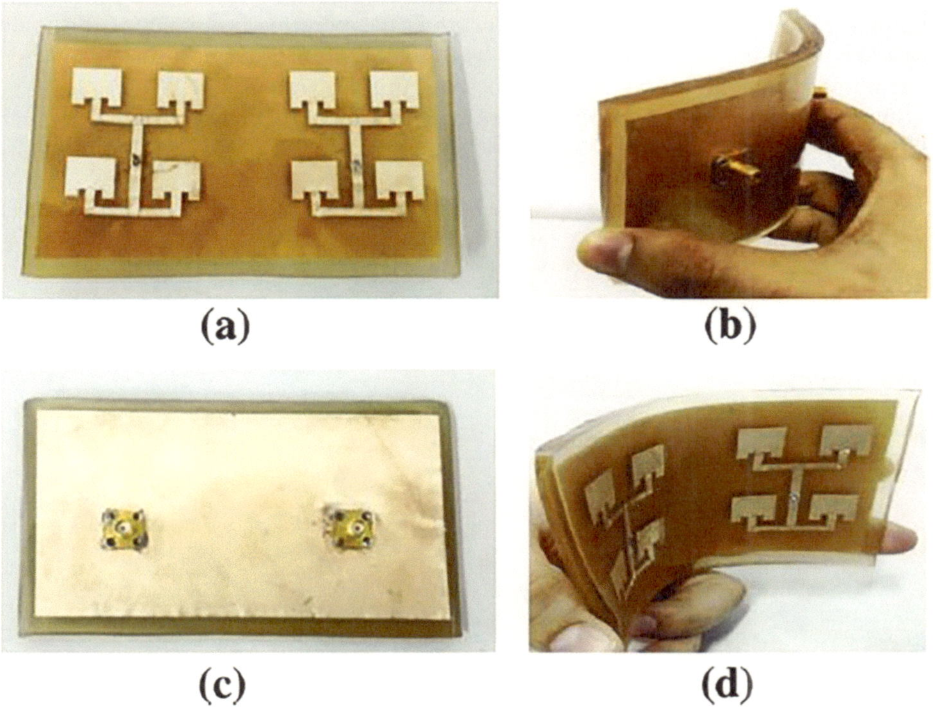

Figure 14.
(a) Front view; (b) bending view; (c) rear view, and (d) front bending view for 2x4 MIMO antenna array [142].

Vaseem *et al*. in 2018 developed Fe_3O_4 nanoparticles based magnetic ink and screen-printed nanocomposite substrates for tunable radio frequency devices [31]. The Fe_3O_4 nanoparticles were synthesized using co-precipitation method. Then, prepared Fe_3O_4 nanoparticles were mixed with SU8 polymeric resin, and the oleic acid was used for the functionalization of Fe_3O_4 nanoparticles in the SU8 resin matrix for the compatibility. The Fe_3O_4 nanoparticles composite was printed by a manual screen-printing technique over a FR4 microwave laminate. **Figure 15** shows the fabrication steps for the freestanding magnetic substrate and antenna [31]. To evaluate the electromagnetic properties of the nanocomposite, a tunable antenna was fabricated on the screen-printed magnetic composite substrate. Frequency tuning of the fabricated antenna in response to the application of magnetostatic fields was successfully demonstrated. For center frequency of 8 GHz, 12.5% tuning was achieved under a magnetic field strength of 3.7 kOe.

Recently, Menezes *et al*. in 2020 fabricated magneto-dielectric bio-composite Fe_3O_4 nanoparticles based flexible film for antenna devices [120]. The flexible film was developed using biopolymeric matrices chitosan (Ch), cellulose (BC) and collagen (Col). The thermal, dielectric, and magnetic properties of flexible film and their application as antenna were tested by fabricating a microstrip patch antenna. For Ch, BC and Col based Fe_3O_4 nanoparticles film, dielectric properties were measured in the range of 5.2–8.3, 6.7–8.4 and 5.9–9.1, respectively from 0 to 5 GHz frequency range. The resonance frequency shifted from 4.66 GHz to 5.89 GHz for different weight percentages of Fe_3O_4 nanoparticles in different polymer matrix. For example, the largest tunability in resonance frequency from 5.55 to 4.69 GHz was observed, by increasing Fe_3O_4 nanoparticles amount from 0 to 80% in Ch matrix with return loss lower than 10 dB for all antenna measurements. The enhancement in bandwidth from 3.37 to 6.34% was reported for all antennas. It was demonstrated in the report that the operating frequency of the devices, size and bandwidth can be modulated by varying the substrate composition, and by controlling the magnetic and dielectric losses of the substrate.

Similar works have been reported by Ghaffar *et al*. in 2018 [143], Cannamela *et al*. 2020 [144], Alqadami *et al*. in 2018 [34] Caprile *et al*. in 2012 [35] by fabricating and

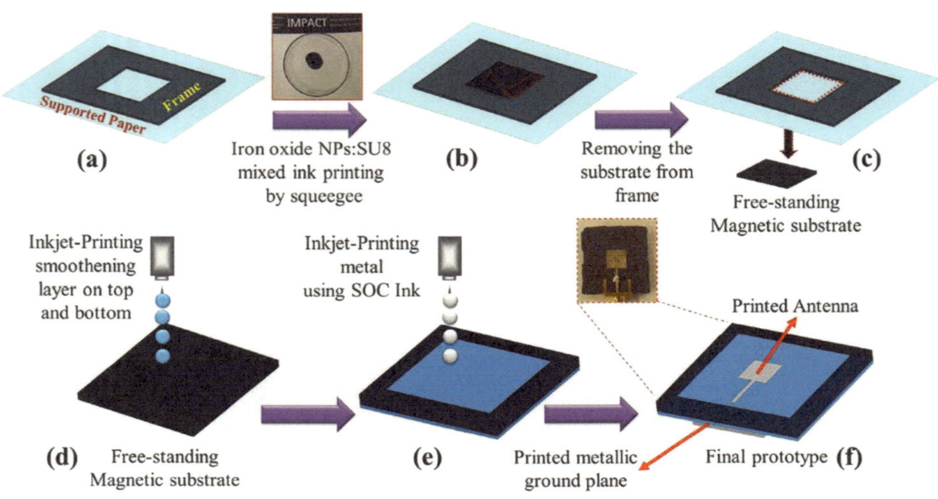

Figure 15.
Step-by-step fabrication process flow of magnetic substrate and printed antenna [31].

characterization of frequency tunable patch antennas using printable inks loaded with Fe_3O_4 nanoparticles.

4.4 Magnetite (Fe_3O_4) nanoparticles composite for circulators and inductors in RF and microwave regions

Ferrites and as magnetite nanoparticle composites have also been used extensively in RF and microwave applications like inductive component, isolators, or as circulators [145, 146]. These devices in electronic industry highly depend on the magnetic properties of the material used. The applications based on soft magnetic ferrite materials take advantage of the fact that spin rotation of these materials changes with the direction of external magnetic field. For one direction, ferrites will absorb the microwave field, and for opposite direction it will transmit the field. This non-reciprocal behavior is the basis of devices such as isolators and circulators [20]. Mostly, Ni-Zn and Mn-Zn ferrites are commonly used for such applications, since they are capable

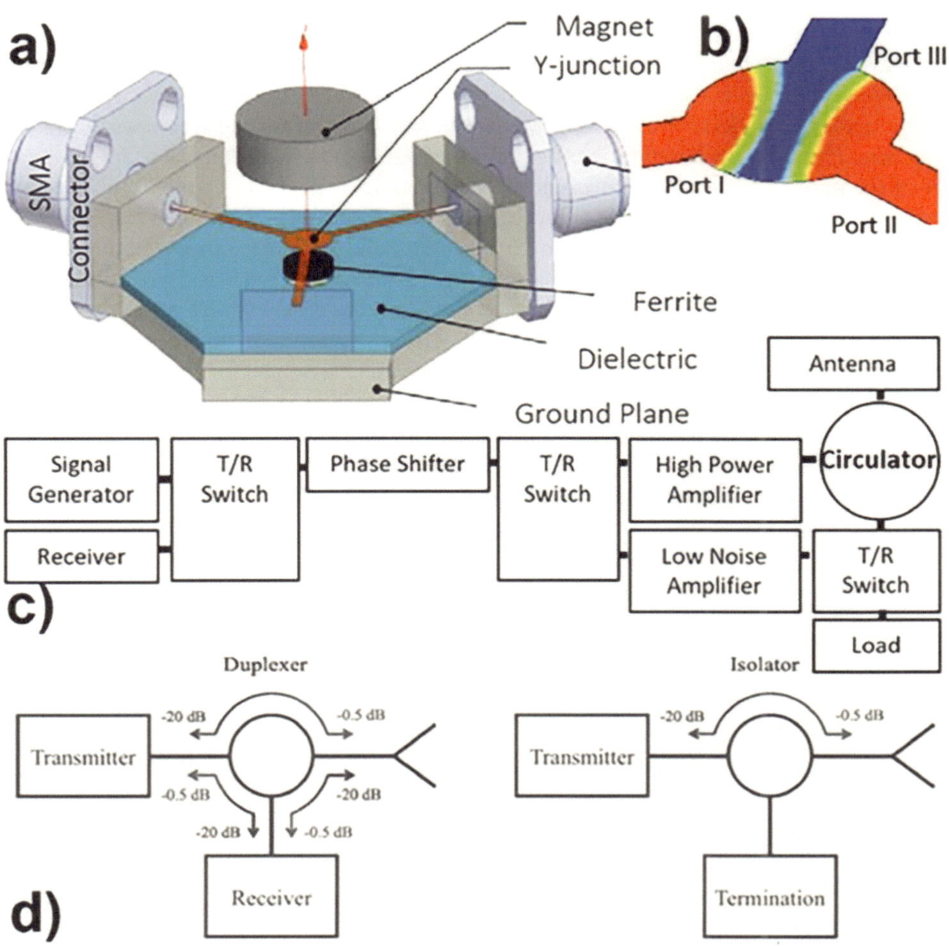

Figure 16.
(a) Design of circulator with use of ferrite in it; (b) electromagnetic simulation of circulator; (c) use of circulator in receiver and transmitter module; and (d) circulator as duplexer and isolator [148].

of providing high permeability, low magnetic loss tangent, high stability, and high resistivity. Nevertheless, they typically exhibit high magnetic losses at higher operating frequencies.

Fe_3O_4 nanoparticles based soft magnetic ferrites can be used for non-reciprocal device applications (e.g., isolators and circulators), because Fe_3O_4 nanoparticles with well controlled particle sizes can offer low magnetic and dielectric losses due to their superparamagnetic property at room temperature. In 2017, Sahasrabudhe *et al.* designed the wideband lumped element circulator based on Fe_3O_4 ferrite and reported 125% improvement in terms of bandwidth at 915 MHz center frequency [147]. There is still limited research on the use of Fe_3O_4 nanoparticles for circulator and inductor applications. **Figure 16** shows use of ferrite material for the design and implementation of circulator along with location and functionality of a circulator within a front-end transmit and receiver module, respectively [148].

5. Conclusion

The chapter presents a review of the key synthesis techniques for magnetite (Fe_3O_4) nanoparticles and their applications. Fe_3O_4 nanoparticles have a large area of applications in different fields such as magnetic separation, storage, biomedical applications, catalyst, water purification, electronics, and so on. It was concluded from the synthesis methods that their structural and magnetic properties are highly dependent on the shape and size of the nanoparticles. The morphology of the particles can be controlled by different synthesis parameters. Among the chemical methods, chemical co-precipitation method is the most advantageous due to the ease of the synthesis approach. Improvement in the stability of Fe_3O_4 nanoparticles with appropriate agents is also discussed in the article. With this regard, the current applications of Fe_3O_4 nanoparticles for RF and microwave applications have been discussed. It is important to tune and tailor control suitable particle size with optimized synthesis approach and applied field strength for the design of RF/ microwave devices and other applications like hyperthermia and drug delivery. For future application of Fe_3O_4 nanoparticles in biomedical device and electronics applications, it is crucial to not only control the morphology and magnetic properties of the nanoparticle but also optimize synthesis methods to increase the yield on industrial scale. Though there are limited studies presently, applications of Fe_3O_4 nanoparticles in RF/Microwave devices is an emerging area, where new application will be discovered in near future. This will open up new avenues in many sectors including biomedical devices.

Author details

Poonam Lathiya[1,2*] and Jing Wang[1,2*]

1 Department of Electrical Engineering, University of South Florida, Tampa, FL, USA

2 Wireless and Microwave Information Systems and RF MEMS Transducers Lab, USA

*Address all correspondence to: poonam2@usf.edu and jingw@usf.edu

References

[1] Samrot AV et al. A review on synthesis, characterization and potential biological applications of superparamagnetic iron oxide nanoparticles. Current Research in Green. 2021;**4**:100042

[2] Bazard P et al. Nanoparticle-based plasmonic transduction for modulation of electrically excitable cells. Scientific Reports. 2017;**7**(1):1-13

[3] Dúran N. Nanotecnologia: introdução, preparação e caracterização de nanomateriais e exemplos de aplicação. Artliber; 2006

[4] Huber DL. Synthesis, properties, and applications of iron nanoparticles. Small. 2005;**1**(5):482-501

[5] Salata OV. Applications of nanoparticles in biology and medicine. Journal of Nanobiotechnology. 2004;**2**(1):1-6

[6] Campos EA et al. Synthesis, characterization and applications of iron oxide nanoparticles-a short review. Journal of Aerospace Technology. 2015;**7**:267-276

[7] Damnjanovic R et al. Hybrid electro-plasmonic neural stimulation with visible-light-sensitive gold nanoparticles. ACS Nano. 2020;**14**(9):10917-10928

[8] Chaturvedi S, Dave PN, Shah NJ. Applications of nano-catalyst in new era. Journal of Saudi Chemical Society. 2012;**16**(3):307-325

[9] Yetter RA, Risha GA, Son SF. Metal particle combustion and nanotechnology. Proceedings of the Combustion Institute. 2009;**32**(2):1819-1838

[10] Manish S et al. Investigation on size dependent structural and magnetic behaviour of nickel ferrite nanoparticles prepared by sol–gel and hydrothermal methods. Materials Chemistry. 2009;**118**(1):174-180

[11] Gupta AK, Gupta M. Synthesis and surface engineering of iron oxide nanoparticles for biomedical applications. Biomaterials. 2005;**26**(18):3995-4021

[12] Issa B et al. Magnetic nanoparticles: Surface effects and properties related to biomedicine applications. International Journal of Molecular Sciences. 2013;**14**(11):21266-21305

[13] Revia RA, Zhang M. Magnetite nanoparticles for cancer diagnosis, treatment, and treatment monitoring: Recent advances. Materials Today. 2016;**19**(3):157-168

[14] Chen L, Bazylinski DA, Lower B. Bacteria that synthesize nano-sized compasses to navigate using earth's geomagnetic field. Nature Education Knowledge. 2010;**3**(10):30

[15] Teja AS, Koh P-Y. Synthesis, properties, and applications of magnetic iron oxide nanoparticles. Progress in Crystal Growth Characterization of Materials. 2009;**55**(1-2):22-45

[16] Wang C-T, Ro S-H. Nanocluster iron oxide-silica aerogel catalysts for methanol partial oxidation. Applied Catalysis A: General. 2005;**285**(1-2):196-204

[17] Jolivet J-P, Tronc E, Chanéac C. Iron oxides: From molecular clusters to solid. A nice example of chemical versatility. Comptes Rendus Geoscience. 2006;**338**(6-7):488-497

[18] Guo H, Barnard AS. Naturally occurring iron oxide nanoparticles:

Morphology, surface chemistry and environmental stability. Journal of Materials Chemistry A. 2013;**1**(1):27-42

[19] Ghazanfari MR et al. Perspective of Fe_3O_4 nanoparticles role in biomedical applications. Biochemistry Research International. 2016;**2016**:32. DOI: 10.1155/2016/7840161

[20] Valenzuela R, Novel applications of ferrites. Physics Research International. 2012;**2012**:9. DOI: 10.1155/2012/591839

[21] Hasany F et al. Magnetic iron oxide nanoparticles: Chemical synthesis and applications review. Current Nanoscience. 2013;**9**(5):561-575

[22] Machala L, Tucek J, Zboril R. Polymorphous transformations of nanometric iron (III) oxide: A review. Chemistry of Materials. 2011;**23**(14):3255-3272

[23] Cornell RM, Schwertmann U. The Iron Oxides: Structure, Properties, Reactions, Occurrences, and Uses. Vol. 2. Federal Republic of Germany: Wiley-vch Weinheim; 2003

[24] Babes L et al. Synthesis of iron oxide nanoparticles used as MRI contrast agents: A parametric study. Journal of Colloid Interface Science. 1999;**212**(2):474-482

[25] Ali A et al. Synthesis, characterization, applications, and challenges of iron oxide nanoparticles. Nanotechnology, Science Applications. 2016;**9**:49

[26] Sun S-N et al. Magnetic iron oxide nanoparticles: Synthesis and surface coating techniques for biomedical applications. Chinese Physics B. 2014;**23**(3):037503

[27] Phan HT, Haes AJ. What does nanoparticle stability mean? The Journal of Physical Chemistry C. 2019;**123**(27):16495-16507

[28] Xu L et al. Stability and reactivity: Positive and negative aspects for nanoparticle processing. Chemical Reviews. 2018;**118**(7):3209-3250

[29] Leonel AG, Mansur AA, Mansur HS. Advanced functional nanostructures based on magnetic iron oxide nanomaterials for water remediation: A review. Water Research. 2021;**190**:116693

[30] Boles J et al. Relationship selling behaviors: antecedents and relationship with performance. Journal of Business & Industrial Marketing. 2000;**15**:141-153

[31] Vaseem M et al. Iron oxide nanoparticle-based magnetic ink development for fully printed tunable radio-frequency devices. Advanced Materials Technologies. 2018;**3**(4):1700242

[32] Lee H et al. Inkjet printed high-Q RF inductors on paper substrate with ferromagnetic nanomaterial. IEEE Microwave Wireless Components Letters. 2016;**26**(6):419-421

[33] Tan L-R et al. Magnetically tunable ferrite loaded SIW antenna. IEEE Antennas Wireless Propagation Letters. 2013;**12**:273-275

[34] Alqadami AS et al. Fabrication and characterization of flexible polymer iron oxide composite substrate for the imaging antennas of wearable head imaging systems. IEEE Antennas Wireless Propagation Letters. 2018;**17**(8):1364-1368

[35] Caprile A et al. Microwave behavior of polymer bonded iron oxide nanoparticles. IEEE Transactions on Magnetics. 2012;**48**(11):3394-3397

[36] Besenhard MO et al. Co-precipitation synthesis of stable iron oxide nanoparticles with NaOH: New insights and continuous production via flow chemistry. Chemical Engineering Journal. 2020;**399**:125740

[37] Ahn T et al. Formation pathways of magnetite nanoparticles by coprecipitation method. The Journal of Physical Chemistry C. 2012;**116**(10):6069-6076

[38] Lee S-J et al. Synthesis and characterization of superparamagnetic maghemite nanoparticles prepared by coprecipitation technique. Journal of Magnetism Magnetic Materials. 2004;**282**:147-150

[39] Araújo-Neto R et al. Monodisperse sodium oleate coated magnetite high susceptibility nanoparticles for hyperthermia applications. Journal of Magnetism Magnetic Materials. 2014;**364**:72-79

[40] Cheraghipour E, Javadpour S, and Mehdizadeh AR. Citrate capped superparamagnetic iron oxide nanoparticles used for hyperthermia therapy. Journal of Biomedical Science and Engineering. 2012;**5**:715-719

[41] Feng B et al. Synthesis of Fe_3O_4/APTES/PEG diacid functionalized magnetic nanoparticles for MR imaging. Colloids Surfaces A: Physicochemical Engineering Aspects. 2008;**328**(1-3):52-59

[42] Sanaeifar N et al. A novel electrochemical biosensor based on Fe_3O_4 nanoparticles-polyvinyl alcohol composite for sensitive detection of glucose. Analytical Biochemistry. 2017;**519**:19-26

[43] Radoń A et al. Structure and optical properties of Fe_3O_4 nanoparticles synthesized by co-precipitation method with different organic modifiers. Materials Characterization. 2017;**131**:148-156

[44] Anbarasu M et al. Synthesis and characterization of polyethylene glycol (PEG) coated Fe_3O_4 nanoparticles by chemical co-precipitation method for biomedical applications. Spectrochimica Acta Part A: Molecular Biomolecular Spectroscopy. 2015;**135**:536-539

[45] Saragi T et al. The impact of synthesis temperature on magnetite nanoparticles size synthesized by co-precipitation method. Journal of Physics: Conference Series. 2018;**1013**(1):012190

[46] Radoń A et al. Electrical conduction mechanism and dielectric properties of spherical shaped Fe_3O_4 nanoparticles synthesized by co-precipitation method. Materials. 2018;**11**(5):735

[47] Sun S, Zeng HJ. Size-controlled synthesis of magnetite nanoparticles. Journal of the American Chemical Society. 2002;**124**(28):8204-8205

[48] Wetterskog E et al. Precise control over shape and size of iron oxide nanocrystals suitable for assembly into ordered particle arrays. Science Technology of Advanced Materials. 2014;**15**(5):055010

[49] Kovalenko MV et al. Fatty acid salts as stabilizers in size-and shape-controlled nanocrystal synthesis: The case of inverse spinel iron oxide. Journal of the American Chemical Society. 2007;**129**(20):6352-6353

[50] Shavel A et al. Synthesis and characterization of iron/iron oxide core/shell nanocubes. Advanced Functional Materials. 2007;**17**(18):3870-3876

[51] Bodnarchuk MI et al. Exchange-coupled Bimagnetic Wüstite/metal

ferrite Core/Shell nanocrystals: Size, shape, and compositional control. Small. 2009;**5**(20):2247-2252

[52] Shavel A, Liz-Marzán LM. Shape control of iron oxide nanoparticles. Physical Chemistry Chemical Physics. 2009;**11**(19):3762-3766

[53] Li D et al. An easy fabrication of monodisperse oleic acid-coated Fe_3O_4 nanoparticles. Materials Letters. 2010;**64**(22):2462-2464

[54] Wooding A et al. Studies of the double surfactant layer stabilization of water-based magnetic fluids. Journal of Colloid Interface Science. 1991;**144**(1):236-242

[55] Zhao SY et al. Synthesis of magnetic nanoparticles of Fe 3 O 4 and CoFe 2 O 4 and their surface modification by surfactant adsorption. Bulletin of the Korean Chemical Society. 2006;**27**(2):237-242

[56] Wang Y et al. One-pot reaction to synthesize superparamagnetic iron oxide nanoparticles by adding phenol as reducing agent and stabilizer. Journal of Nanoparticle Research. 2012;**14**(4):1-7

[57] Wu W, He Q, Jiang C. Magnetic iron oxide nanoparticles: Synthesis and surface functionalization strategies. Nanoscale Research Letters. 2008;**3**(11):397-415

[58] Vuong TKO et al. Synthesis of high-magnetization and monodisperse Fe_3O_4 nanoparticles via thermal decomposition. Materials Chemistry Physics. 2015;**163**:537-544

[59] Joseyphus RJ et al. Size controlled Fe nanoparticles through polyol process and their magnetic properties. Materials Chemistry Physics. 2010;**123**(2-3):487-493

[60] Abbas M et al. Synthesis of high magnetization hydrophilic magnetite (Fe_3O_4) nanoparticles in single reaction—Surfactantless polyol process. Ceramics International. 2013;**39**(7):7605-7611

[61] Cai W, Wan JJ. Facile synthesis of superparamagnetic magnetite nanoparticles in liquid polyols. Journal of Colloid Interface Science. 2007;**305**(2):366-370

[62] Liu H-L et al. One-pot polyol synthesis of monosize PVP-coated sub-5 nm Fe_3O_4 nanoparticles for biomedical applications. Journal of Magnetism Magnetic Materials. 2007;**310**(2):e815-e817

[63] Gonçalves RH, Cardoso CA, Leite ER. Synthesis of colloidal magnetite nanocrystals using high molecular weight solvent. Journal of Materials Chemistry. 2010;**20**(6):1167-1172

[64] Maity D et al. Polyol-based synthesis of hydrophilic magnetite nanoparticles. Journal of Applied Physics. 2010;**107**(9):09B310

[65] Cheng C, Xu F, Gu H. Facile synthesis and morphology evolution of magnetic iron oxide nanoparticles in different polyol processes. New Journal of Chemistry. 2011;**35**(5):1072-1079

[66] Bunge A et al. Correlation between synthesis parameters and properties of magnetite clusters prepared by solvothermal polyol method. Journal of Materials Science. 2019;**54**(4):2853-2875

[67] Xuan S et al. Tuning the grain size and particle size of superparamagnetic Fe_3O_4 microparticles. Chemistry of Materials. 2009;**21**(21):5079-5087

[68] Wang W et al. Size-controlled synthesis of water-dispersible superparamagnetic Fe_3O_4 nanoclusters

and their magnetic responsiveness. RSC Advances. 2015;**5**(92):75292-75299

[69] Choi K-H et al. A facile fabrication of Fe_3O_4/ZnO core-shell submicron particles with controlled size. IEEE Transactions on Magnetics. 2011;**47**(10):3369-3372

[70] Zhuang L et al. Hydrophilic magnetochromatic nanoparticles with controllable sizes and super-high magnetization for visualization of magnetic field intensity. Scientific Reports. 2015;**5**(1):1-9

[71] Oh AH et al. Synthesis of Fe_3O_4 nanoparticles of various size via the polyol method. Ceramics International. 2020;**46**(8):10723-10728

[72] Sayed FN, Polshettiwar V. Facile and sustainable synthesis of shaped iron oxide nanoparticles: Effect of iron precursor salts on the shapes of iron oxides. Scientific Reports. 2015;**5**(1):1-14

[73] Polshettiwar V et al. High-surface-area silica nanospheres (KCC-1) with a fibrous morphology. Angewandte Chemie International Edition. 2010;**49**(50):9652-9656

[74] Gan YX et al. Hydrothermal Synthesis of Nanomaterials. Hindawi: Journal of Nanomaterials; 2020;**2020**:3

[75] Torres-Gómez N. et al. Shape tuning of magnetite nanoparticles obtained by hydrothermal synthesis: Effect of temperature. Journal of Nanomaterials. 2019:**2019**:15

[76] Sun X et al. Size-controlled synthesis of magnetite (Fe_3O_4) nanoparticles coated with glucose and gluconic acid from a single Fe (III) precursor by a sucrose bifunctional hydrothermal method. The Journal of Physical Chemistry C. 2009;**113**(36):16002-16008

[77] Xu Z et al. Oleylamine as both reducing agent and stabilizer in a facile synthesis of magnetite nanoparticles. Chemistry of Materials. 2009;**21**(9):1778-1780

[78] Hou L et al. Shape-controlled nanostructured magnetite-type materials as highly efficient Fenton catalysts. Applied Catalysis B: Environmental. 2014;**144**:739-749

[79] Li Y et al. Single-microemulsion-based solvothermal synthesis of magnetite microflowers. Ceramics International. 2014;**40**(3):4791-4795

[80] Sato J et al. Hydrothermal synthesis of magnetite particles with uncommon crystal facets. Journal of Asian Ceramic Societies. 2014;**2**(3):258-262

[81] Lopez Perez J et al. Advances in the preparation of magnetic nanoparticles by the microemulsion method. The Journal of Physical Chemistry B. 1997;**101**(41):8045-8047

[82] Liu Z et al. Synthesis of magnetite nanoparticles in W/O microemulsion. Journal of Materials Science. 2004;**39**(7):2633-2636

[83] Bumajdad A et al. Characterization of iron hydroxide/oxide nanoparticles prepared in microemulsions stabilized with cationic/non-ionic surfactant mixtures. Journal of Colloid Interface Science 2011;**355**(2):282-292

[84] Maleki H et al. Size-controlled synthesis of superparamagnetic iron oxide nanoparticles and their surface coating by gold for biomedical applications. Journal of Magnetism Magnetic Materials. 2012;**324**(23):3997-4005

[85] Lu T et al. Surfactant effects on the microstructures of Fe_3O_4 nanoparticles

synthesized by microemulsion method. Colloids Surfaces A: Physicochemical Engineering Aspects. 2013;**436**:675-683

[86] Asab G, Zereffa EA, and Abdo Seghne T. Synthesis of silica-coated Fe_3O_4 nanoparticles by microemulsion method: Characterization and evaluation of antimicrobial activity. International Journal of Biomaterials. 2020;**2020**:11

[87] Liu W et al. One step-synthesis of monodispersed and fluorescent mesoporous mSiO2-coated Fe_3O_4 nanoparticles. Glass Physics Chemistry. 2021;**47**(6):590-600

[88] Teng Y et al. Magnetic iron oxide nanoparticle-hollow mesoporous silica spheres: Fabrication and potential application in drug delivery. Current Applied Physics. 2020;**20**(2):320-325

[89] Gonzalez-Rodriguez R, Campbell E, Naumov A. Multifunctional graphene oxide/iron oxide nanoparticles for magnetic targeted drug delivery dual magnetic resonance/fluorescence imaging and cancer sensing. PLoS One. 2019;**14**(6):e0217072

[90] Asgari M et al. A robust method for fabrication of monodisperse magnetic mesoporous silica nanoparticles with core-shell structure as anticancer drug carriers. Journal of Molecular Liquids. 2019;**292**:111367

[91] Simeonidis K et al. Controlling magnetization reversal and hyperthermia efficiency in core–shell iron–iron oxide magnetic nanoparticles by tuning the interphase coupling. ACS Applied Nano Materials. 2020;**3**(5):4465-4476

[92] Hedayatnasab Z et al. Polycaprolactone-coated superparamagnetic iron oxide nanoparticles for in vitro magnetic hyperthermia therapy of cancer. European Polymer Journal. 2020;**133**:109789

[93] Dadfar SM et al. Size-isolation of superparamagnetic iron oxide nanoparticles improves MRI, MPI and hyperthermia performance. Journal of Nanobiotechnology. 2020;**18**(1):1-13

[94] Ranoo S et al. Enhancement in hyperthermia efficiency under in situ orientation of superparamagnetic iron oxide nanoparticles in dispersions. Applied Physics Letters. 2019;**115**(4):043102

[95] Gholipour N et al. Development of Ga-68 labeled, biotinylated thiosemicarbazone dextran-coated iron oxide nanoparticles as multimodal PET/MRI probe. International Journal of Biological Macromolecules. 2020;**148**:932-941

[96] Ansari MO et al. Superparamagnetic iron oxide nanoparticles based cancer theranostics: A double edge sword to fight against cancer. Journal of Drug Delivery Science Technology. 2018;**45**:177-183

[97] Khoshneviszadeh M et al. Evaluation of surface-modified superparamagnetic iron oxide nanoparticles to optimize bacterial immobilization for bio-separation with the least inhibitory effect on microorganism activity. Nanoscience Nanotechnology-Asia. 2020;**10**(2):166-174

[98] Adams SA et al. Fe_3O_4@ SiO2 nanoparticles functionalized with gold and poly (vinylpyrrolidone) for bio-separation and sensing applications. ACS Applied Nano Materials. 2018;**1**(3):1406-1412

[99] Samrot AV et al. A study on influence of superparamagnetic iron oxide nanoparticles (SPIONs) on green gram (Vigna radiata L.) and earthworm

(Eudrilus eugeniae L.). Materials Research Express. 2020;7(5):055002

[100] Singh AD et al. Impact of cobalt oxide nanoparticles on the morpho-physiological and biochemical response in plants. In: Plant Responses to Nanomaterials. Springer Nature. Switzerland AG: Springer; 2021. pp. 249-267

[101] Ahmed B et al. Interplay between engineered nanomaterials (ENMs) and edible plants: A current perspective. In: Phytotoxicity of Nanoparticles. Springer International Publishing AG; 2018. pp. 63-102

[102] Elazab HA. Hydrothermal synthesis of graphene supported Pd/Fe_3O_4 nanoparticles as efficient magnetic catalysts for Suzuki Cross–Coupling. Chemical Engineering. 2019:8

[103] Wu W et al. Sensitive, selective and simultaneous electrochemical detection of multiple heavy metals in environment and food using a lowcost Fe_3O_4 nanoparticles/fluorinated multi-walled carbon nanotubes sensor. Ecotoxicology Environmental Safety. 2019;**175**:243-250

[104] Zhang L et al. Magnetic solid-phase extraction based on Fe_3O_4/graphene oxide nanoparticles for the determination of malachite green and crystal violet in environmental water samples by HPLC. International Journal of Environmental Analytical Chemistry. 2018;**98**(3):215-228

[105] Bahiraei M, Hangi M, Rahbari A. A two-phase simulation of convective heat transfer characteristics of water–Fe_3O_4 ferrofluid in a square channel under the effect of permanent magnet. Applied Thermal Engineering. 2019;**147**:991-997

[106] Barakat M et al. Facile synthesis of muscovite–supported Fe_3O_4 nanoparticles as an adsorbent and heterogeneous catalyst for effective removal of methyl orange: Characterisation, modelling, and mechanism. Journal of the Taiwan Institute of Chemical Engineers. 2021;**119**:146-157

[107] Veisi H et al. Palladium nanoparticles anchored polydopamine-coated graphene oxide/Fe_3O_4 nanoparticles (GO/Fe_3O_4@ PDA/Pd) as a novel recyclable heterogeneous catalyst in the facile cyanation of haloarenes using K4 [Fe (CN) 6] as cyanide source. Journal of Industrial Engineering Chemistry. 2020;**90**:379-388

[108] Sohouli E et al. Electrochemical sensor based on modified methylcellulose by graphene oxide and Fe_3O_4 nanoparticles: Application in the analysis of uric acid content in urine. Journal of Electroanalytical Chemistry. 2020;**877**:114503

[109] Basheer AA. Advances in the smart materials applications in the aerospace industries. Aircraft Engineering Aerospace Technology. 2020;**92**(7):1027-1035

[110] Chatterjee S et al. Design of multifunctional fluorescent hybrid materials based on SiO2 materials and core–shell Fe_3O_4@ SiO2 nanoparticles for metal ion sensing. Small. 2019;**15**(44):1904569

[111] Hashim A, Agool IR, Kadhim KJ. Novel of (polymer blend-Fe_3O_4) magnetic nanocomposites: Preparation and characterization for thermal energy storage and release, gamma ray shielding, antibacterial activity and humidity sensors applications. Journal of Materials Science: Materials in Electronics. 2018;**29**(12):10369-10394

[112] Du L et al. A fascinating metallo-supramolecular polymer network with thermal/magnetic/light-responsive

shape-memory effects anchored by Fe_3O_4 nanoparticles. Macromolecules. 2018;**51**(3):705-715

[113] Ansari MS et al. Reactively sputtered half-metallic Fe_3O_4 thin films at room temperature on polymethyl methacrylate: A perspective for flexible spintronics. Ceramics International. 2020;**46**(11):19302-19310

[114] Martynenko IV et al. Magneto-fluorescent microbeads for bacteria detection constructed from superparamagnetic Fe_3O_4 nanoparticles and AIS/ZnS quantum dots. Analytical Chemistry. 2019;**91**(20):12661-12669

[115] Lathiya P, Wang J. Effects of the sintering temperature on RF complex permeability of NiCuCoZn ferrites for near-field communication applications. IEEE Transactions on Magnetics. 2018;**55**(2):1-4

[116] Vural M et al. Nanostructured flexible magneto-dielectrics for radio frequency applications. Journal of Materials Chemistry C. 2014;**2**(4):756-763

[117] Yang G-M et al. Tunable miniaturized patch antennas with self-biased multilayer magnetic films. IEEE Transactions on Antennas Propagation. 2009;**57**(7):2190-2193

[118] Feng S, et al. Hysteresis loss reduction in self bias FeSi/$SrFe_{12}O_{19}$ soft magnetic composites. Chinese Physics B. 2021;**31**:027503

[119] Wu T et al. The low-frequency improvement with loading soft magnetic ferrite films for multiband antenna applications. International Journal of Applied Electromagnetics Mechanics. 2021;**66**(2):359-368

[120] de Menezes FL et al. From magneto-dielectric biocomposite films to microstrip antenna devices. Journal of Composites Science. 2020;**4**(4):144

[121] Kuanr BK et al. Size dependent microwave properties of ferrite nanoparticles: Application to microwave devices. Journal of Applied Physics. 2009;**105**(7):07B522

[122] Jadav M, Bhatnagar S. Particle size controlled magnetic loss in magnetite nanoparticles in RF-microwave region. IEEE Transactions on Magnetics. 2020;**56**(7):1-8

[123] Ganesan V et al. Size-controlled synthesis of superparamagnetic magnetite nanoclusters for heat generation in an alternating magnetic field. Journal of Molecular Liquids. 2019;**281**:315-323

[124] Raikher YL, Stepanov VI. Ferromagnetic resonance in a suspension of single-domain particles. Physical Review B. 1994;**50**(9):6250

[125] Lalan V, Ganesanpotti SJ. Broadband electromagnetic response and enhanced microwave absorption in carbon black and magnetic Fe_3O_4 nanoparticles reinforced polyvinylidenefluoride composites. Journal of Electronic Materials. 2020;**49**(3):1666-1676

[126] Zhu L et al. Hydrothermal synthesis of magnetic Fe_3O_4/graphene composites with good electromagnetic microwave absorbing performances. Journal of Magnetism Magnetic Material. 2017;**426**:114-120

[127] Li H et al. Preparation and characterization of highly water-soluble magnetic Fe_3O_4 nanoparticles via surface double-layered self-assembly method of sodium alpha-olefin sulfonate. Journal of Magnetism Magnetic Materials. 2015;**384**:213-218

[128] Qiao M et al. Application of yolk–shell Fe_3O_4@ N-doped carbon nanochains as highly effective microwave-absorption material. Nano Research. 2018;**11**(3):1500-1519

[129] Song X et al. Study on large-scale spatial dynamic absorption of Fe_3O_4@ SiO2 Shell–Core and Nano-Fe_3O_4 magnetic particles. ACS Applied Electronic Materials. 2021;**3**(11):5066-5076

[130] Lathiya P, Kreuzer M, Wang J. RF complex permeability spectra of Ni-Cu-Zn ferrites prepared under different applied hydraulic pressures and durations for wireless power transfer (WPT) applications. Journal of Magnetism Magnetic Materials. 2020;**499**:166273

[131] Sun Q et al. Activated carbon fiber/Fe 3 O 4 composite with enhanced electromagnetic wave absorption properties. RSC Advances. 2018;**8**(61):35337-35342

[132] Zhou P et al. Microwave absorption properties of SiC@ SiO2@ Fe_3O_4 hybrids in the 2-18 GHz range. International Journal of Minerals, Metallurgy, Materials. 2017;**24**(7):804-813

[133] Lathiya P, Wang J. Enhancement in magnetic permeability of Ni-Co-Zn ferrites using CuO doping for RF and microwave devices. Journal of the American Ceramic Society. 2022;**105**(4):2678-2689

[134] Hosseini SH, Asadnia AJ. Polyaniline/Fe_3O_4 coated on MnFe2O4 nanocomposite: Preparation, characterization, and applications in microwave absorption. International Journal of Physical Sciences. 2013;**8**(22):1209-1217

[135] Xu Y et al. Preparation of a reduced graphene oxide/SiO_2/Fe_3O 4 UV-curing material and its excellent microwave absorption properties. RSC Advances. 2017;**7**(29):18172-18177

[136] Wang Z et al. Enhanced microwave absorption of Fe 3 O 4 nanocrystals after heterogeneously growing with ZnO nanoshell. RSC Advances. 2013;**3**(10):3309-3315

[137] Yang R-B et al. Synthesis and characterization of Fe_3O_4/polypyrrole/carbon nanotube composites with tunable microwave absorption properties: Role of carbon nanotube and polypyrrole content. Chemical Engineering Journal. 2016;**285**:497-507

[138] Liu X et al. Enhanced microwave absorption properties in GHz range of Fe_3O_4/C composite materials. Journal of Alloys Compounds. 2015;**649**:537-543

[139] Liu J et al. Microwave absorption enhancement of multifunctional composite microspheres with spinel Fe_3O_4 cores and anatase TiO_2 shells. Small. 2012;**8**(8):1214-1221

[140] Morales CA. Magneto-Dielectric Polymer Nanocomposite Engineered Substrate for RF and Microwave Antennas. University of South Florida. ProQuest Dissertations Publishing. 2011:3482482

[141] Castro J et al. Synthesis and characterization of low-loss Fe 3 O 4-PDMS magneto-dielectric polymer nanocomposites for RF applications. In: WAMICON 2014. Tampa, Fl. USA: IEEE; 2014. pp. 1-5. DOI: 10.1109/WAMICON.2014.6857768

[142] Alqadami ASM et al. Polymer (PDMS-Fe_3O_4) magneto-dielectric substrate for a MIMO antenna array. Applied Physics A. 2016;**122**(1):1-7

[143] Ghaffar FA et al. Design and fabrication of a frequency and

polarization reconfigurable microwave antenna on a printed partially magnetized ferrite substrate. IEEE Transactions on Antennas Propagation. 2018;**66**(9):4866-4871

[144] Cannamela M et al. Fabrication of magnetic polymer nanocomposites using inkjet 3D print technology. In: NIP & Digital Fabrication Conference. Vol. 2020, No. 1. Society for Imaging Science and Technology; 2020. pp. 1-5

[145] Zaquine I, Benazizi H, Mage JJ. Ferrite thin films for microwave applications. Journal of Applied Physics. 1988;**64**(10):5822-5824

[146] Batoo KM, Ansari MS. Low temperature-fired Ni-Cu-Zn ferrite nanoparticles through auto-combustion method for multilayer chip inductor applications. Nanoscale Research Letters. 2012;**7**(1):1-14

[147] Sahasrabudhe M et al. Design of wideband lumped element circulator. In: 2017 Progress in Electromagnetics Research Symposium-Spring (PIERS). IEEE; 2017. pp. 3602-3608

[148] Harris VG, Sokolov AS, Magnetism N. The self-biased circulator: Ferrite materials design and process considerations. Journal of Superconductivity Novel Magnetism. 2019;**32**(1):97-108

Chapter 6

Application of Iron Oxide in Supercapacitor

Rajan Lakra, Rahul Kumar, Parasanta Kumar Sahoo, Sandeep Kumar and Ankur Soam

Abstract

Iron oxide nanostructures have been considered very promising material as electrode in electrochemical energy storage devices because of their lower cost of synthesis and high theoretical charge storage capacity. Iron oxide nanoparticles and their nanocomposites have performed excellent in supercapacitor. Iron oxide as negative electrode has extended the working voltage window of a supercapacitor. The main problems associated with iron oxide based electrodes are their poor electrical conductivity and cycle stability. Therefore, a conductive carbon matrix has been added to the iron oxide based electrodes to improve the electrochemical performance. In this chapter, recent progress on iron oxide and its composite with different materials as electrode in supercapacitor is summarized. The various synergistic effects of nanocomposites and compositional engineering to enhance the electrochemical performance of iron oxide are also discussed.

Keywords: iron oxide, composite, electrodes, supercapacitor

1. Introduction

Serious environmental and climate problems have been arisen due to the continuous consumption and depletion of fossil fuels [1–4]. This problem would be worst in the coming days as the demand of energy is going increased day by day. Consequently, there is a need of clean and sustainable energy resources such as solar cells and batteries. The solar cell always needs sun light to generate continuous electricity supply. This problem can be sort out by integrating an energy storage device to the solar cell. Batteries can be used for the above work however their replacement and maintenance are major issues [1, 3, 5–7]. Supercapacitors have drawn much attention in recent years to be used in several applications such as industrial power and energy management. Moreover, they are very efficient (fast charging/discharging and long life), low-cost and environmentally friendly [8–11]. They have larger power density than batteries and capability to deliver the charge quickly [12–15]. A supercapacitor is fabricated with two electrodes separated by a separator, and electrolyte (**Figure 1**). The electrode materials have decisive role in the charge storage capacity; therefore, different materials and their combinations have been used in supercapacitor [2, 16–19]. Supercapacitor can store the energy via two process (1)

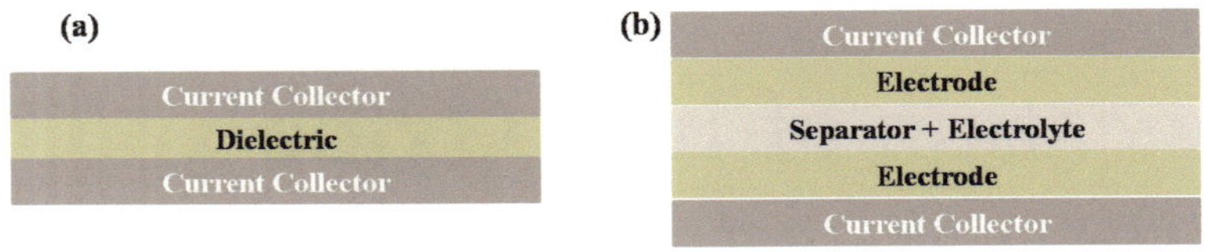

Figure 1.
Structure of (a) dielectric capacitor and (b) supercapacitor.

formation of electric double layer (EDL) at the electrode/electrolyte interface [4, 19, 20] and (2) redox reactions (pseudocapacitor) [16, 21, 22] (**Figure 2**). Carbon and silicon based materials are generally used for EDL capacitor whereas pseudocapacitor consists of metal oxides and conducting polymers. A hybrid supercapacitor technology has also been developed by combining the properties of EDLC and pseudocapacitor [24–27].

Carbon based electrodes can offer high power and long cycle stability. However, low energy density is major concern with carbon electrodes [28, 29]. In the past decade, major efforts have been made developing a hybrid supercapacitor in order to meet the demands of high energy and high power density [30, 31]. Nano-sized structure of metal oxides can offer higher surface area to the electrode for good electrode/electrolyte ions interaction and short ion-diffusion length. Among metals oxides, iron oxides (FeO_x) are being considered efficient electrodes as they have low cost of synthesis, high theoretical capacitance due to variable oxidation states [32–34]. Various nanostructures of Fe based materials such as nanorods, nanosheets, nanotubes and nanoparticles, nanowires, nanoflowers etc. have been employed to increase the surface area and shorten the diffusion length of ions [2, 21, 32, 35]. The above structures have shown excellent performance in supercapacitor. However, low conductivity and poor structural stability are major issues with iron oxides, which need to be addressed. Incorporating conductive materials to iron oxides can lead to improve the conductivity and stability as well. Therefore, different forms of carbon and polymers have been mixed with iron oxide to fabricate a hybrid electrode and this electrode has shown excellent performance in supercapacitor. In this chapter, the achievements and progress made towards developing iron oxides (Fe_2O_3 and Fe_3O_4) based electrodes have been reviewed.

2. Supercapacitor

A supercapacitor or ultracapacitor is an advanced technology which has much larger capacitance than a normal capacitor. Supercapacitors can store the energy about 10 times more than dielectric capacitors and return the energy back faster than batteries. They occupy the place between dielectric capacitors and rechargeable batteries in Ragone plot (**Figure 2a**). The charge storage process in supercapacitor is completely different than that of conventional capacitor. In electrostatic capacitor, two conducting electrodes are separated by a dielectric material. The charge is stored by the polarization process in the dielectric material. In case of supercapacitor, an electrolyte is filled between two electrodes of high surface area. During charging, opposite charge starts accumulating on the surface of electrodes. As the charge is stored physically with no chemical change, it is highly reversible and it shows stable

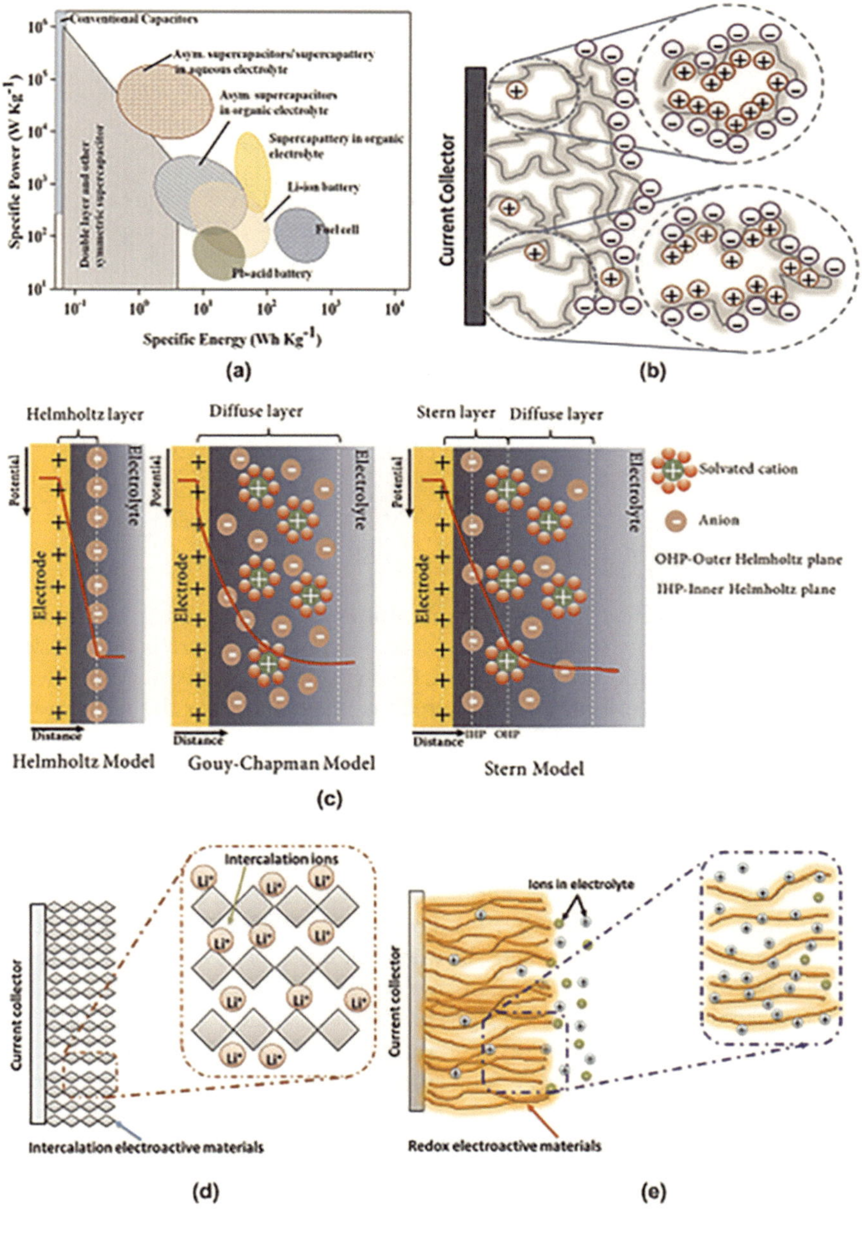

Figure 2.
(a) Ragone plot for different devices. (b) Schematic representation of charge storage at high surface area electrode. (c) Different models for EDL: Helmholtz model, Gouy–Chapman model and Gouy–Chapman–Stern model. Charge storage process via redox process (pseudocapacitor); (d) bulk redox and (e) surface redox. (Reprinted with permission from Ref. [23] Copyright (2020) American Chemical Society).

behavior for a large number of cycles. Capacitance (C) is defined as the ratio of stored charge (Q) to the applied voltage (V)

$$C = \frac{Q}{V} \tag{1}$$

C is directly proportional to the surface area (A) of each electrode and inversely proportional to the distance (d) between the electrodes

$$C = \frac{\varepsilon A}{d} \tag{2}$$

Where ε is the permittivity of the insulating material, A is the area of the electrode and d is the distance between the electrodes. The energy (E) of a capacitor is calculated by the following equation;

$$E = \frac{1}{2}CV^2 \tag{3}$$

Power (P) is the energy delivered by a device per unit time. The maximum power is limited by equivalent series resistance (ESR) with the following relation

$$P = \frac{V^2}{4 * ESR} \tag{4}$$

An EDLC has two electrodes which are immersed in an electrolyte and separated by a separator. Its charge storage is non faradaic in nature, there is no transfer of charge between electrode and electrolyte. Electric charge is accumulated electrostatically on electrode/electrolyte interface which is responsible for the formation of EDL (**Figure 2b**). The charging and discharging processes are highly reversible and there is no chemically change in the electrodes. This leads to the increased in cycle life, compared to batteries but less than conventional capacitor.

Pseudo means false so, as the name indicates this type of materials do not store charge like a conventional capacitor that is by forming EDL but work like a capacitor. The charge storage mechanism of such type of capacitor is faradaic in nature or the electrode reacts with the electrolyte to store charge (**Figure 2d** and **e**). Transition metal oxide and polymer show such kind of behavior. A reversible redox processes take place in which the valence electrons of electro active materials are transferred across the electrode/electrolyte interface, resulting in a potential-dependent capacitance (**Figure 2e**). The term pseudo-capacitance is used to distinguish this mechanism of charge storage from double layer. So we can say that that pseudo capacitance possess battery-like behavior with faradic reactions between electrodes and electrolyte. **Table 1** shows the difference between EDLC and pseudocapacitor.

EDL capacitor	Pseudocapacitor
It stores charge by non-faradaic mechanism.	It stores charge by faradaic mechanism.
Capacitance is low as compared to pseudo capacitance.	Capacitance is 10 times higher than EDLC.
Large cycle stability	Poor cycle stability
The charging/discharging curve is highly reversible	Reversible but not as compared to EDLC

Table 1.
Comparison between EDLC and pseudocapacitor.

3. Fe_2O_3 based electrodes

Fe_2O_3 electrode stores the charge via reversible oxidation/reduction reactions at the electrode and electrolyte interface. Low conductivity and poor cycling stability hamper its application in supercapacitor. Shivakumara et al. [36] have used porous α-Fe_2O_3 as electrode in supercapacitor. Porous α-Fe_2O_3 synthesized by sol–gel route offered BET surface area of 386 $m^2\ g^{-1}$. α-Fe_2O_3 demonstrated specific capacitance of 300 F g^{-1} in 0.5 M Na_2SO_3 at a discharge rate of 1 A g^{-1}. However, the electrode could retain only 73% of the initial capacitance after 1000 cycles.

Flower-like α-Fe_2O_3 nanostructures prepared by ethylene glycol mediated self-assembly process exhibited capacitance value of 127 F g^{-1} in 0.5 M Na_2SO_3 determined at a current density of 1 A g^{-1} [37]. Flower-like α-Fe_2O_3 has retained 80% of the initial capacitance after 1000 cycles. Despite of good value of capacitance, Fe_2O_3 does not meet the requirement of long cycle life. Change in volume during multiple cycles causes large degradation in the capacitance. Therefore, maximum efforts have been made towards shorting the diffusion length of electrolyte ions and minimizing the volume change by combining Fe_2O_3 with conducting materials.

Large efforts have been made to add a conductive phase to Fe_2O_3 electrodes to enhance overall capacitive performance. Abad et al. [38] have synthesized rGO-Fe_2O_3 nanocomposites by hydrothermal process for supercapacitor application. In the composite electrode, Fe_2O_3 nanoparticles have average size of 25 nm and are anchored on graphene sheets making good physical contact. The Fe_2O_3 nanoparticles prevent the restacking of rGO sheets which results in good accessibility of electrolyte ions to the electrode. rGO-Fe_2O_3 exhibited a specific capacitance of 291 F g^{-1} at 1 A g^{-1} in 1 M KOH aqueous solution. The performance was observed stable in the negative potential window of −1 to 0 V with three-electrode system. Fe_2O_3 worked well as negative electrode in asymmetric supercapacitor. A hybrid structure of porous Fe_2O_3 nanosheets on carbon fabric (CF-Fe_2O_3) was developed in order to overcome poor electrical conductivity and poor cycling stability of Fe_2O_3 [39]. Fe_2O_3 was deposited on CF via electrodeposition process. An asymmetric supercapacitor with CF-Fe_2O_3 as negative electrode and CF-Co_3O_4 as positive electrode was fabricated, which exhibited a good value of areal capacitance to be 842 mF cm^{-2}. The asymmetric supercapacitor demonstrated maximum volumetric energy density of 6.75 mWh cm^{-3} and power density of 104 mW cm^{-3}. The device retained 93% capacitance after 4000 cycles.

Jiang et al. [40] have synthesized Fe_2O_3/porous carbon nanocomposites (Fe_2O_3/HAC) by hydrothermal method. The porous carbon prevents the agglomeration of Fe_2O_3 nanoparticles and minimizes the ion diffusion path on the electrode surface, providing more flow channels for the ions. Fe_2O_3/HAC demonstrated larger capacitance of 156 F g^{-1} than that of individual components of HAC (145 F g^{-1}) and Fe_2O_3 (90 F g^{-1}). Redox peaks were observed in CV of both Fe_2O_3 and Fe_2O_3/HAC. The obtained CV consisted of a pair of oxidation peaks at −0.652 and −0.641 V and reduction peaks at −0.996 and −1.038 V, respectively. Fe_2O_3/HAC stores the charge via following mechanism;

$$FeOOH + H_2O + e- \rightarrow Fe(OH)_2 + OH^{-1} \tag{5}$$

$$Fe(OH)_2 + OH^{-1} \rightarrow FeOOH + H_2O + e^- \tag{6}$$

A specific capacitance of 970 F g^{-1} was achieved from Fe_2O_3/rGO composite which was prepared by in situ synthesis and mechanical agitation methods [41]. This high value of capacitance is mainly due to the synergistic effect between Fe_2O_3 and rGO.

rGO provides high electrical conductivity to the composite electrode and makes fast channels for the movement of charges. A degradation of 25% of capacitance was observed after 2000 cycles. A carbon shell layer has been coated on porous Fe_2O_3 nanowire arrays by hydrothermal route to improve the electrochemical properties of the electrode [42]. Fe_2O_3/C was used as the anode in solid state asymmetric supercapacitor with MnO_2 as the cathode. A wide voltage of 2 V was achieved with this configuration. The fabricated device could give a maximum energy density of 30.625 Wh kg^{-1} and a maximum power density of 5000 W kg^{-1}. The device exhibited good stability about 82% retention of capacitance after 10,000 cycles. A α-Fe_2O_3/MnO_x and α-Fe_2O_3/C core-shell nanostructures were developed to design an asymmetric supercapacitor [43]. In the hybrid electrode, fast reversible redox-reactions occur on MnO_x and α-Fe_2O_3 NR core facilitates electron transfer between shell and the current collector. Similarly, α-Fe_2O_3/C core-shell works as negative electrode. The asymmetric device fabricated with α-Fe_2O_3/C//α-Fe_2O_3/MnO_x core-shell exhibited specific capacitance of 1.28 F cm^{-3} at a scan rate of 10 mV s^{-1} with voltage window of 0–2 V. The supercapacitor was able to retain 78% capacitance at the scan rate of 400 mVs^{-1} with a maximum energy-density of ~0.64 mWh cm^{-3} and a maximum power-density of 155 mW cm^{-3}. Zhang et al. [44] have developed Fe_2O_3@NiO nanorods on flexible carbon cloth by hydrothermal method. A specific capacitance of 800 mF cm^{-2} at 10 mA cm^{-2} was obtained with Fe_2O_3@NiO/CC electrode. Remarkable long cycle stability (96.8% capacitance retention after 16,000 cycles) was demonstrated by the electrode. This outstanding performance from Fe_2O_3@NiO/CC is due to the good electrical contact of active material with flexible carbon.

α-Fe_2O_3 has shown good electrochemical performance with graphitic (g-C_3N_4) carbon nitride [45]. The as-prepared g-C_3N_4/α-Fe_2O_3 composites with large specific surface area exhibited specific capacitance of 580 F g^{-1} determined at 1.0 A g^{-1} in 1M KOH. The Fe_2O_3 nanoparticles accommodate the space between the layers of g-C_3N_4 avoiding restacking and increasing the probability of interaction between Fe_2O_3 and electrolyte. Zhong et al. [46] have fabricated a heterostructure of Fe_2O_3 nanospheres anchored on FeS_2 nanosheets by one-step hydrothermal approach for supercapacitor. The FeS_2@Fe_2O_3 hybrid electrode demonstrated a specific capacitance of 255 F g^{-1} as well as good cycle stability, 90% capacitance retention after 5000 cycles. The excellent long cycle life could be ascribed to hybrid structure which can facilitate fast transportation of the electrolyte ions. A hybrid electrode of PEDOT coated onto Ti-doped Fe_2O_3 showed a remarkable capacitive performance [47]. A high value of areal capacitance of 1.15 F cm^{-2} at 1 mA cm^{-2} has been achieved with this strategy. Ti-Fe_2O_3@PEDOT demonstrated extraordinary cyclic stability of 96% retention in capacitance after 30,000 cycles, which is better than that of Ti-Fe_2O_3 (80.7%) and Fe_2O_3 (81.8%) electrodes. Ti-doping in the electrode enhances electrical conductivity and better utilization of Fe_2O_3. PEDOT has two important roles here, working as protective layer for Fe_2O_3 and relaxing the transfer of electrons.

A high specific capacitance of 1124 F g^{-1} has been achieved form a ternary composite of polyaniline/Fe_2O_3 decorated graphene coated on carbon cloth (**Figures 3** and **4**) at a current of 0.25 A g^{-1} in 1 M H_2SO_4 [48]. This large value of specific capacitance is due to the synergistic effects of individual component in the electrode. Graphene contributes as EDL capacitance and Fe_2O_3 and polyaniline as pseudocapacitance to the overall capacitance of hybrid electrode. In-situ polymerization of polyaniline also leads to improve the surface area of electrodes. Le et al. [50] have fabricated a hybrid electrode of polypyrrole-coated Fe_2O_3 nanotubes (Fe_2O_3@PPy) for high-performance electrode for aqueous asymmetric supercapacitor. A thin layer of polypyrrole was coated on Fe_2O_3

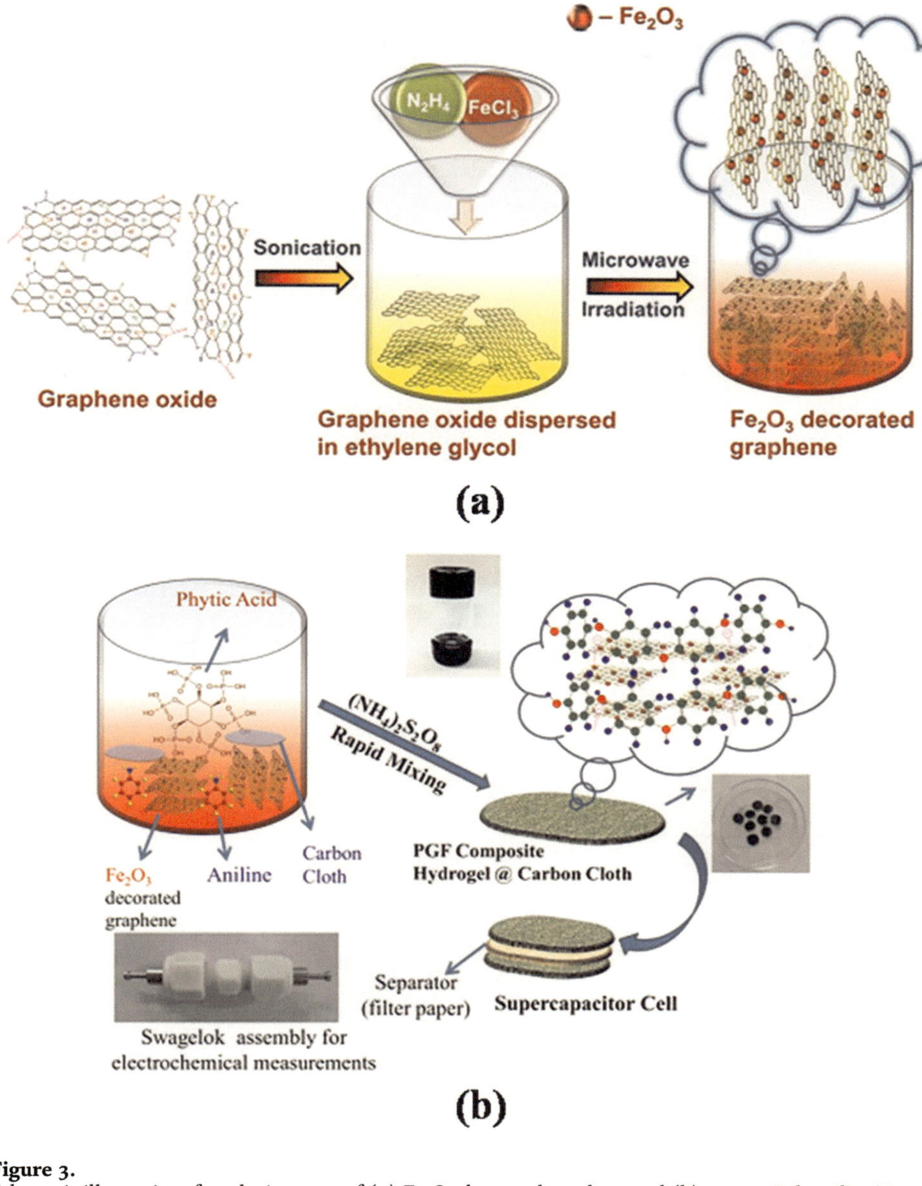

Figure 3.
Schematic illustration of synthesis process of (a) Fe_2O_3 decorated graphene and (b) ternary Polyaniline/Graphene/Fe_2O_3 (PGF) composite on carbon cloth. (Reprinted with permission from Ref. [48]. Copyright (2020) American Chemical Society).

nanotubes by an in situ chemical oxidative polymerization method. Fe_2O_3@PPy could deliver a capacitance of 530 mF cm^{-2} at 1 mA cm^{-2}. There is synergistic effect between PPy and Fe_2O_3, the shell of conducting PPy makes an ease transportation for charge carriers and improves stability of Fe_2O_3 nanotubes while charging/discharging process. The fabricated asymmetric supercapacitor with Fe_2O_3@PPy and MnO_2 nanotubes could deliver energy density of 51.2 Wh kg^{-1} at a power density of 285.4 W kg^{-1} operated at high voltage of 2.0 V. 83.5% capacitance retention was observed over 5000 cycles with the asymmetric supercapacitor.

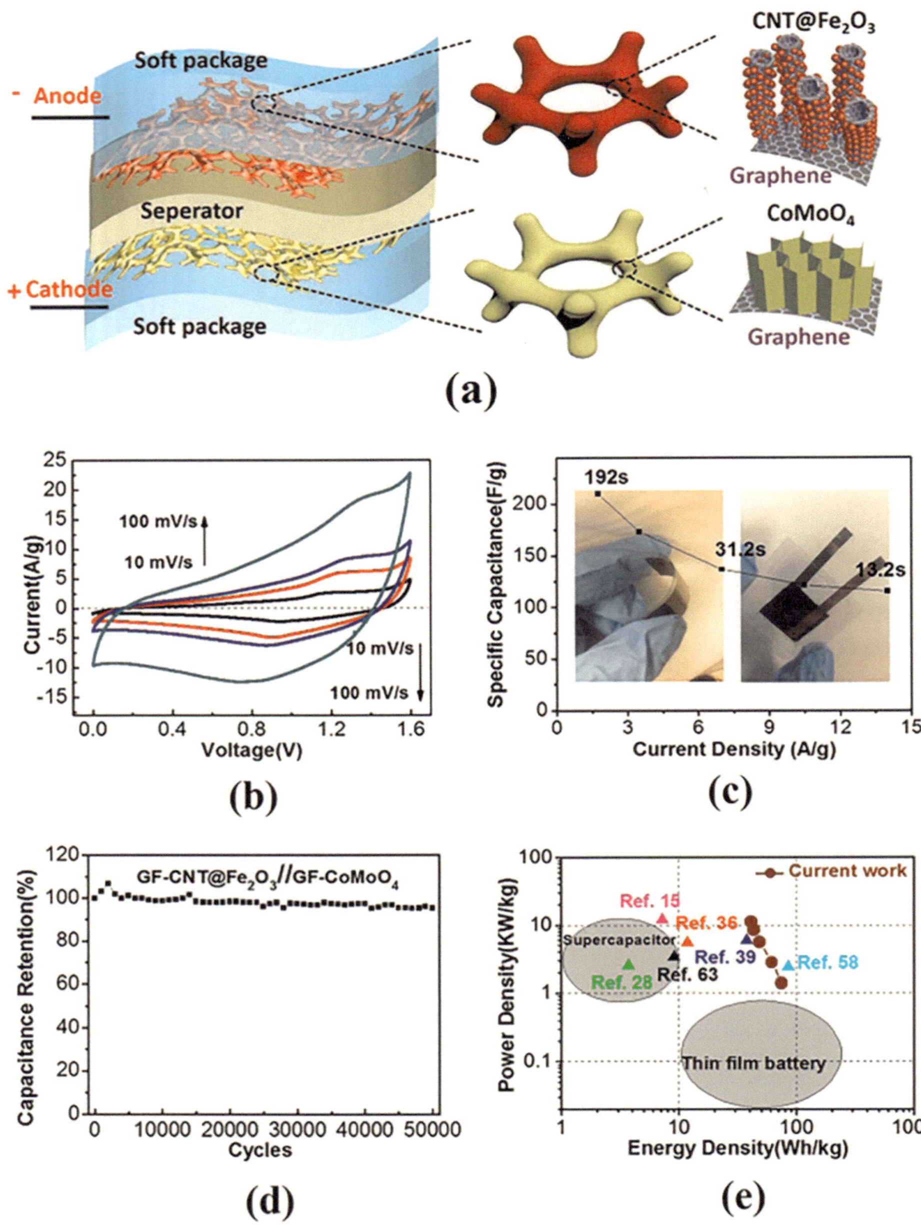

Figure 4.
Electrochemical behavior of the GF-CNT@Fe_2O_3//GF-$CoMoO_4$ supercapacitor. (a) Full-cell structure. (b) CV curves, (c) rate capability, (d) cycling life, (e) Ragone plot of the asymmetric supercapacitor. Digital photos of the single electrode and full cell are depicted in insets of (c). (Reprinted with permission from Ref. [49]. Copyright (2015) American Chemical Society).

Guan et al. [49] have developed an ultrahigh-energy and long-life supercapacitor anode material by coating a thin layer of Fe_2O_3 on graphite foam-carbon nanotube framework GF-CNT@Fe_2O_3. GF-CNT@Fe_2O_3 as anode was also integrated into a full supercapacitor cell with GF-$CoMoO_4$ as cathode. A high energy density of $\sim$74.7 Wh kg^{-1} with power density of$\sim$1400 W kg^{-1} has been obtained from the full device. The device also exhibited exceptional cycle stability, retention of about 95.4% capacitance after 50,000 cycles. These values make it a promising candidate for high performance supercapacitor. Improvement in the electrochemical performance of Fe_2O_3 based electrode may also be possible by introducing nitrogen in the electrode. In this regard, Zhao et al. [51] have synthesized nitrogen-doped graphene and Fe_2O_3 composites (NGFeCs) by hydrothermal method. NGFeCs exhibited improved electrochemical performance than GFeCs. The specific capacitance of NGFeCs electrode was determined to be 260.1 F g^{-1} (150.4 Fg^{-1} for GFeCs electrode) with current density of 2 A g^{-1}. In addition to above, 82.5% retention in capacitance was observed after 1000 cycles. The improvement in the performance is due to the good electronic conductivity and increased active sites.

Self-assembled α-Fe_2O_3 mesocrystals/graphene nanohybrids demonstrated a good value of specific capacitance of 306.9 F g^{-1} at 3 A g^{-1} in an aqueous electrolyte and in voltage window of 1 V [33]. Porous structure of α-Fe_2O_3 mesocrystals/graphene nanohybrids with high electrical conductivity can provide fast transportation for electrolyte ions even at high discharge current densities. 2D α-Fe_2O_3/rGO nanocomposites fabricated by one-pot solvothermal approach exhibited specific capacitance value to be 903 F g^{-1} calculated at a current density of 1 A g^{-1} [52]. This value of capacitance was observed superior than α-Fe_2O_3 nanoplates. In the hybrid composite, α-Fe_2O_3 nanoplates were encapsulated in rGO nanosheets to form a porous structure. The α-Fe_2O_3/rGO nanocomposites with high electrical conductivity and 2D nanostructure promote the charge transportation between electrode and electrolyte, hence enhancing the electrochemical performance of the electrode.

4. Fe_3O_4 based electrodes

Transition metal oxides (RuO_2, MnO_2, NiO, Fe_2O_3, Co_3O_4, etc.) have been used as electrode materials for supercapacitor application because they use rapid reversible redox reactions at the surface of active materials to offer high power density [9, 31, 53–58]. Amongst the metal oxides, Fe_3O_4 is a better alternative electrode material for supercapacitor because of its low cost, natural abundance, environmental friendliness, high theoretical specific capacitance (2299 F g^{-1}), large potential window and different valence states [59–61]. It has been seen in literature that Fe_3O_4 with various dimensions and morphology shows high value of specific capacitance [62].

The charge storage mechanism of Fe_3O_4 was investigated in various aqueous electrolytes such as sodium chloride, potassium hydroxide, sodium sulphate, sodium sulphite and sodium phosphate. Fe_3O_4 in the Na_2SO_3 electrolyte demonstrated a higher capacitance of 510 F g^{-1} with an operating potential of 1.2 V and this value was greater than those of other electrolytes [63]. Two pseudocapacitive mechanisms were purposed for Fe_3O_4 in the Na_2SO_3 electrolyte, which are given below

$$FeO + SO_3^{2-} \leftrightarrow FeSO_4 + 2e^- \tag{7}$$

$$2Fe^{II}O + SO_3^{2-} \leftrightarrow \left(Fe^{III}O\right) + SO_3^{2-}\left(Fe^{III}O\right) + 2e^- \tag{8}$$

Eq. (7) represents the surface redox reaction of sulphur in the form of sulphate and sulphite anions while Eq. (8) represents the oxidation/reduction reaction between Fe (II) and Fe(III). It was observed that due to the addition of both the EDLC and pseudocapacitor, the capacitance of Fe_3O_4 in the Na_2SO_3 electrolyte rises and involves the reduction/oxidation of specially adsorbed sulphite anions as given by the following equations [63].

$$2SO_3^{2-} + 3H_2O + 4e^- \leftrightarrow S_2O_3^{2-} + 6OH^- \quad (9)$$

$$S_2O_3^{2-} + 3H_2O + 8e^- \leftrightarrow 2S^{2-} + 6OH^- \quad (10)$$

Many methods have been used to fabricate the Fe_3O_4 materials for supercapacitor application such as hydrothermal method [64–67], Electroplating [63], sol-gel method [68], Chemical precipitation method [69], Hydrolysis and annealing process [70], solvothermal method [71], Electrospinning [72], Aerosol method [73]. Reports on different synthesis processes of Fe_3O_4 based electrodes and their specific capacitance values are given in **Table 2**.

Fe_3O_4@carbon nanosheet composite synthesized using ammonium ferric citrate precursor and graphene oxide as the structure-directing agent under a hydrothermal process was used as electrode in supercapacitor [59]. Fe_3O_4@carbon nanosheet composite exhibited a specific capacitance of 586 F g^{-1} at 0.5 A g^{-1} in KOH/PVA gel electrolyte. This composite showed excellent energy density of 18.3 Wh kg^{-1} and power density of 351 W kg^{-1}.

Wang et al. [60] developed a simple route to fabricate the hybrid electrode of Fe_3O_4-doped porous carbon nanorods (Fe_3O_4-DCN) supported by three dimensional (3D) kenaf stem-derived macroporous carbon (KSPC) for supercapacitor. The 3D-KSPC/Fe_3O_4-DCN showed excellent specific capacitance of 285.4 F g^{-1} at the

Electrode	Synthesis process	Electrolyte	Capacitance	Ref.
Fe_2O_3/porous carbon	Hydrothermal	6 M KOH	256 F g^{-1} at 1 A g^{-1}	[40]
FeS_2 nanosheet@Fe_2O_3 nanosphere	Hydrothermal	1 M Li_2SO_4	255 F g^{-1} at 1 A g^{-1}	[46]
α-Fe_2O_3 nanostructures	Sol–gel route	0.5 M Na_2SO_3	300 F g^{-1} at 1 A g^{-1}	[36]
Polyaniline/graphene/ Fe_2O_3 composites hydrogel	Microwave irradiation and in-situ polymerization	1 M H_2SO_4	1124 F g^{-1} at 0.25 A g^{-1}	[48]
α-Fe_2O_3/rGO	One-pot solvothermal approach	1 M KOH	903 F g^{-1} at 1 A g^{-1}	[52]
Polypyrrole-coated Fe_2O_3 nanotubes	In situ chemical oxidative polymerization method	1 M Na_2SO_4	530 mF cm^{-2} at 1 mA cm^{-2}	[50]
Porous-Fe_2O_3@C nanowire	Hydrothermal route	1 M Na_2SO_4	280 F g^{-1} at 1 Ag^{-1} and 241.3 mF cm^{-2} at 1 mA cm^{-2}	[42]
α-Fe_2O_3/rGO composites	In situ synthesis and mechanical agitation methods	6 M KOH	970 F g^{-1} at 1 A g^{-1}	[41]
Porous flowerlikeα-Fe_2O_3	Ethylene glycol mediated self-assembly process	0.5 M Na_2SO_3	127 F g^{-1} at 1 A g^{-1}	[37]
Core-branch Fe_2O_3@NiO nanorods	Hydrothermal method	3 M KOH	800 mF cm^{-2} at 10 mA cm^{-2}	[44]

Table 2.
Electrochemical capacitance of some Fe_2O_3 based electrodes.

current density of 1 A g^{-1} and excellent cyclic stability in 2.0 M KOH electrolyte (**Figure 5**). Kumar et al. [61] prepared a hybrid electrode of 3D rGO NSs containing Fe_3O_4 NPs using one-pot microwave approach. The experimental studies showed that the as-synthesized Fe_3O_4/rGO hybrids were made of faceted Fe_3O_4 NPs with

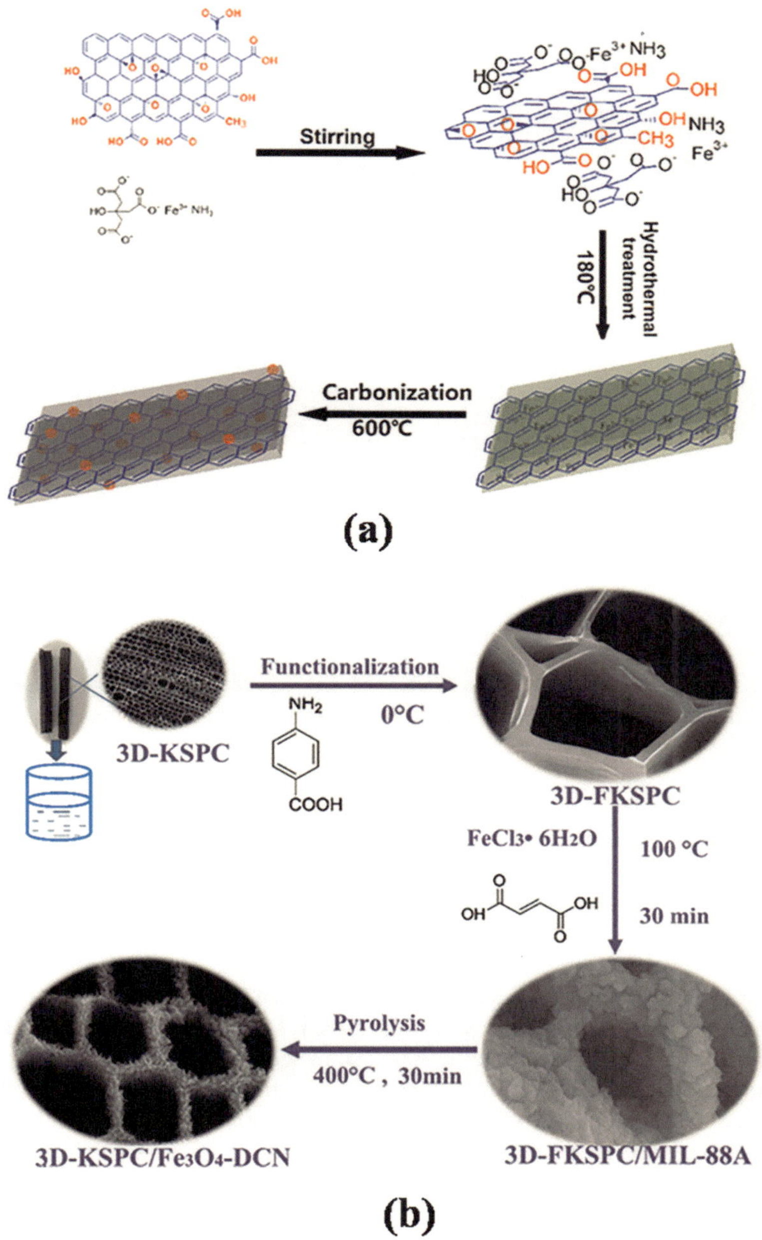

Figure 5.
(a) Schematic illustration of the formation process of carbon nanosheet embedded Fe_3O_4. (Reprinted with permission from Ref. [59]. Copyright (2016) American Chemical Society.) (b) Schematic illustration of the formation process of 3D-KSPC/Fe_3O_4-DCN nanocomposites. (Reprinted with permission from Ref. [60]. Copyright (2016) American Chemical Society).

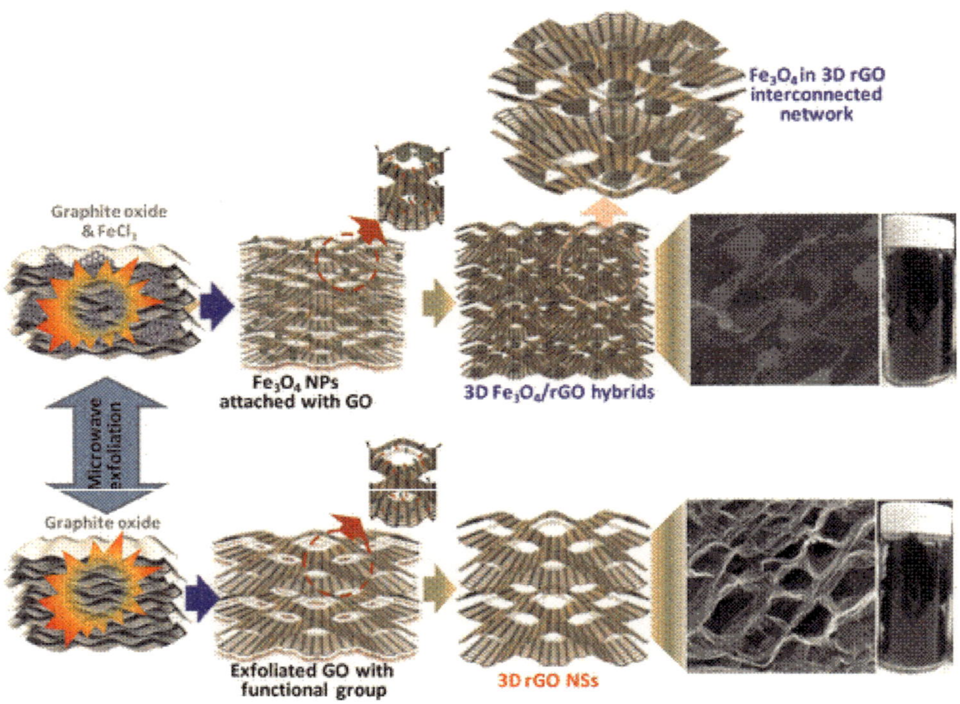

Figure 6.
Schematic formation mechanism of 3D Fe_3O_4/rGO hybrids. (Reprinted with permission from Ref. [61]. Copyright (2017) American Chemical Society).

interconnected network of rGO NSs. The schematic formation mechanism is shown in **Figure 6**. This material as electrode in supercapacitor exhibited a specific capacitance of 455 F g^{-1} at the scan rate of 8 mV s^{-1}.

Chen et al. [74] fabricated a Fe_3O_4 film with regularly edge-affected cubic morphology on stainless foil using hydrothermal method and this film was used as electrode for supercapacitor. This film exhibited a specific capacitance of 118.2 F g^{-1} at the current of 6 mA in Na_2SO_3 electrolyte and good cycle stability about 88.75% capacitance retention after 500 cycles. Wang et al. [63] prepared the Fe_3O_4 film by an electroplating method to be used in supercapacitor as electrode. Fe_3O_4 film demonstrated a specific capacitance of 170 F g^{-1} in 1 M Na_2SO_3 electrolyte. Fe_3O_4 nanoparticles synthesized by sol gel method exhibited a specific capacitance of 185 F g^{-1} at the current of 1 mA in 3 M KOH electrolyte [68]. Brousse et al. [69] prepared Fe_3O_4 by chemical precipitation, which demonstrated a specific capacitance of 75 F g^{-1} in aqueous electrolyte of 0.1 M K_2SO_4. Liu et al. [64] synthesized Fe_3O_4 nanorods and carbon coated Fe_3O_4 nanorods via hydrothermal reaction and subsequent sintering procedure. Fe_3O_4 nanorods showed a specific capacitance of 208.6 F g^{-1} while carbon coated Fe_3O_4 nanorods exhibited a specific capacitance of 275.9 F g^{-1} at a current density of 0.5 A g^{-1} in 1 M Na_2SO_3 aqueous solution.

Guan et al. [65] synthesized carbon nanotube/Fe_3O_4 (CNT/Fe_3O_4) nanocomposite by an easy and efficient hydrothermal method. CNT/Fe_3O_4 nanocomposite showed enhanced specific capacitances of 117.2 F g^{-1} (which is three times than that of pure Fe_3O_4) at current density of 10 mA cm^{-2} in 6 M KOH electrolyte. It also exhibited superior cyclic stability and energy density of

Electrode	Method	Electrolyte	Specific capacitance	Reference
Fe_3O_4 thin film	Electroplating	1 M Na_2SO_3	170 F g^{-1}	[63]
Fe_3O_4 nanoparticles	Sol–gel	3 M KOH	185 F g^{-1}	[68]
Fe_3O_4	Chemical precipitation method	0.1 M K_2SO_4	75 F g^{-1}	[69]
Fe_3O_4 nanorods	Hydrothermal	1 M Na_2SO_3	208.6 F g^{-1}	[64]
Fe_3O_4/graphene	Hydrothermal	1 M KOH	661 F g^{-1}	[66]
Fe_3O_4/reduced graphene oxide	Hydrolysis and annealing process	6 M KOH	350.6 F g^{-1}	[70]
Fe_3O_4/reduced graphene oxide	Solvothermal process	1 M KOH	480 F g^{-1}	[75]
Fe_3O_4–carbon nanosheets	Hydrothermal and heat treatment	1 M Na_2SO_3	163.4 F g^{-1}	[67]
Fe_3O_4 doped double-shelled hollow carbon spheres	Aerosol, spray and in-situ polymerization methods	6 M KOH	1153 F g^{-1}	[76]

Table 3.
Capacitance of Fe_3O_4 based electrodes synthesized by different process.

16.2 Wh kg^{-1}. Lin et al. [66] synthesized Fe_3O_4/GO by using hydrothermal method, which showed a specific capacitance of 661 F g^{-1} at current density of 1 A g^{-1} in 1 M KOH electrolyte. Qi et al. [70] prepared the Fe_3O_4/rGO composites using hydrolysis route and subsequent annealing process for supercapacitor application. The Fe_3O_4/rGO composite showed specific capacitance of 350.6 F g^{-1} at 1 mV s^{-1} in 6 M KOH electrolyte (**Table 3**).

It is seen that iron oxides have demonstrated excellent performance in supercapacitor and asymmetric supercapacitor as well [71–73, 77–79]. Shi et al. [75] prepared the nanocomposites of Fe_3O_4 NPs connected to rGO sheets by using a solvothermal process. Fe_3O_4/rGO nanocomposites were utilized to prepare thin film supercapacitor electrodes by using a spray deposition technique without the addition of insulating binders. It was observed that the Fe_3O_4/rGO nanocomposite exhibited higher specific capacitance, 480 F g^{-1} at a discharge current density of 5 A g^{-1}, which is larger than that of pure rGO or pure Fe_3O_4 NPs. Fe_3O_4/rGO nanocomposite exhibited energy density of 67 W h kg^{-1} and power density of 5506 W kg^{-1}. Ultrathin nanoporous Fe_3O_4–carbon nanosheets synthesized by hydrothermal method demonstrated a specific capacitance of 163.4 F g^{-1} in 1 M Na_2SO_3 electrolyte [67].

Mu et al. [80] synthesized Fe_3O_4 nanosheets on one-dimensional (1D) carbon nanofibers (CNFs) using the electrospinning technique and solvent-thermal process. The Fe_3O_4/CNFs nanocomposite exhibited a specific capacitance of 135 F g^{-1} in 1 M Na_2SO_3 electrolyte. The high capacitance may be due to the improved electrical-conductivity of the composite by adding the CNFs in Fe_3O_4. Li et al. [76] prepared the double-shelled hollow carbon spheres with conductive graphite structure and Fe_3O_4 species (C-C-Fe_3O_4) using aerosol, spray, and in-situ polymerization methods. The C-C-Fe_3O_4 showed remarkable value of specific capacitance to be 1153 F g^{-1} at current density of 2 A g^{-1}. C-C-Fe_3O_4 also demonstrated good rate capability, 514 F g^{-1} at 100 A g^{-1}. The electrode exhibited only 3.3% degradation in the capacitance after 8000 cycles. Energy density of 17–45 W h kg^{-1} and powder density of 400–8000 W kg^{-1} were achieved from the C-C-Fe_3O_4 electrode.

5. Conclusion

Recent progress on supercapacitor is to increase its specific capacitance, rate capability, cycle life and operating voltage by developing low cost and eco-friendly electrode materials. In this regard, iron oxide may be a suitable candidate with higher value of capacitance as negative electrode for asymmetric supercapacitor to be operated at wide voltage window. Therefore, a lot of work has been carried out on iron oxide based electrode for supercapacitor application. The electrochemical performance of iron oxide based electrodes has been reviewed and discussed in this chapter. It is observed that the synthesis process and morphology of iron oxide play important role in supercapacitor performance. Efforts have been made towards preparing a composite of iron oxide with high conductive materials in order to overcome its poor electrical conductivity. Nanostructure of iron oxide such as nanoparticles could enhance the active sites for electrochemical reactions. Iron oxide nanostructure with carbon can further increase the specific capacitance and energy density of supercapacitor. Synthesis of iron oxide nanoparticles/graphene nanocomposite seems to be more effective approach for high performance supercapacitor.

Conflict of interest

The authors declare no conflict of interest.

Author details

Rajan Lakra[1], Rahul Kumar[2], Parasanta Kumar Sahoo[1], Sandeep Kumar[3] and Ankur Soam[1*]

1 Department of Mechanical Engineering, Siksha 'O' Anusandhan University, Bhubaneswar, Odisha, India

2 Centre for Energy Studies, Indian Institute of Technology Delhi, New Delhi, India

3 Department of Electromechanical, Systems and Metal Engineering and Center for Molecular Modeling (CMM), Ghent University, Gent, Belgium

*Address all correspondence to: ankursoam007@gmail.com

References

[1] Kim T, Song W, Son DY, Ono LK, Qi Y. Lithium-ion batteries: Outlook on present, future, and hybridized technologies. Journal of Materials Chemistry A. 2019;**7**(7):2942-2964. DOI: 10.1039/C8TA10513H

[2] Lv L et al. Progress in iron oxides based nanostructures for applications in energy storage. Nanoscale Research Letters. 2021;**16**(1). DOI: 10.1186/s11671-021-03594-z

[3] Nitta N, Wu F, Lee JT, Yushin G. Li-ion battery materials: Present and future. Materials Today. 2015;**18**(5):252-264. DOI: 10.1016/j.mattod.2014.10.040

[4] Lakra R, Kumar R, Sahoo PK, Thatoi D, Soam A. A mini-review: Graphene based composites for supercapacitor application. Inorganic Chemistry Communications. 2021;**133**. DOI: 10.1016/j.inoche.2021.108929

[5] Zhao D, Wang Y, Zhang Y. High-performance Li-ion batteries and super-capacitors base on 1-D nanomaterials in prospect. 2011;**3**(1). DOI: 10.1007/BF03353653

[6] Kim H-J, Krishna TNV, Zeb K, Rajangam V, Gopi CVV, Sambasivam S, et al. A comprehensive review of Li-ion battery materials and their recycling techniques. Electronics. 2020;**9**(7):1161. DOI: 10.3390/electronics9071161

[7] Kuperman A, Aharon I. Battery – Ultracapacitor hybrids for pulsed current loads: A review. Renewable and Sustainable Energy Reviews. 2011;**15**(2): 981-992. DOI: 10.1016/j.rser.2010.11.010

[8] Kumar R, Soam A, Sahajwalla V. Sucrose-derived carbon-coated nickel oxide (SDCC-NiO) as an electrode material for supercapacitor applications. Materials Advances. 2020;**1**(4):609-616. DOI: 10.1039/d0ma00323a

[9] Kumar R, Soam A, Hussain R, Mansuri I, Sahajwalla V. Carbon coated iron oxide (CC-IO) as high performance electrode material for supercapacitor applications. Journal of Energy Storage. 2020;**32**(August):101737. DOI: 10.1016/j.est.2020.101737

[10] Kumar R, Singh M, Soam A. Study on electrochemical properties of silicon micro particles as electrode for supercapacitor application. Surfaces and Interfaces. 2020;**19**. DOI: 10.1016/j.surfin.2020.100524

[11] Soam A, Kumar R, Thatoi D, Singh M. Electrochemical performance and working voltage optimization of nickel ferrite/graphene composite based supercapacitor. Journal of Inorganic and Organometallic Polymers and Materials. 2020;**30**(9):3325-3331. DOI: 10.1007/s10904-020-01540-7

[12] Soam A, Kavle P, Kumbhar A, Dusane RO. Performance enhancement of micro-supercapacitor by coating of graphene on silicon nanowires at room temperature. Current Applied Physics. 2017;**17**(2). DOI: 10.1016/j.cap.2016.11.011

[13] Kumagai S, Mukaiyachi K, Tashima D. Rate and cycle performances of supercapacitors with different electrode thickness using non-aqueous electrolyte. Journal of Energy Storage. 2015;**3**:10-17. DOI: 10.1016/j.est.2015.08.002

[14] Zhi M, Xiang C, Li J, Li M, Wu N. Nanostructured carbon-metal oxide composite electrodes for supercapacitors: A review. Nanoscale. 2013;**5**(1):72-88. DOI: 10.1039/c2nr32040a

[15] Soam A, Mahender C, Kumar R, Singh M. Power performance of BFO-graphene composite electrodes based supercapacitor. Materials Research Express. 2019;**6**(2). DOI: 10.1088/2053-1591/aaf125

[16] Iro ZS, Subramani C, Dash SS. A brief review on electrode materials for supercapacitor. International Journal of Electrochemical Science. 2016; **11**(12):10628-10643. DOI: 10.20964/2016.12.50

[17] Ke Q, Wang J. Graphene-based materials for supercapacitor electrodes – A review. Journal of Materials. 2016;**2** (1):37-54. DOI: 10.1016/j.jmat.2016.01.001

[18] Li Y, Wang H, Mousavi F, Kaner RB. Graphene-based materials for flexible supercapacitors. Chemical Society Reviews. 2015;**44**:3639-3665. DOI: 10.1039/C4CS00316K

[19] Soam A, Arya N, Singh A, Dusane R. Fabrication of silicon nanowires based on-chip micro-supercapacitor. Chemical Physics Letters. 2017;**678**. DOI: 10.1016/j.cplett.2017.04.019

[20] Zhai Y, Dou Y, Zhao D, Fulvio PF, Mayes RT. Carbon Materials for Chemical Capacitive Energy Storage2011. pp. 4828-4850. DOI: 10.1002/adma.201100984

[21] Nithya VD, Arul NS. Review on α-Fe_2O_3 based negative electrode for high performance supercapacitors. Journal of Power Sources. 2016;**327**:297-318. DOI: 10.1016/j.jpowsour.2016.07.033

[22] Gogotsi Y, Penner RM. Energy storage in nanomaterials – Capacitive, pseudocapacitive, or battery-like? ACS Nano. 2018;**12**(3):2081-2083. DOI: 10.1021/acsnano.8b01914

[23] Balasubramaniam S, Mohanty A, Balasingam SK, Kim SJ, Ramadoss A. Comprehensive insight into the mechanism, material selection and performance evaluation of supercapatteries. Nano-Micro Letters. 12, 2020;(1):1-46. DOI: 10.1007/s40820-020-0413-7

[24] H. M. N. Graphene, et al., "High-energy MnO_2 nanowire/graphene and graphene asymmetric electrochemical capacitors," vol. 4, no. 10, pp. 5835–5842, 2010. doi: 10.1021/nn101754k

[25] Sankar KV, Selvan RK. The ternary $MnFe_2O_4$/graphene/polyaniline hybrid composite as negative electrode for supercapacitors. Journal of Power Sources. 2015;**275**:399-407. DOI: 10.1016/j.jpowsour.2014.10.183

[26] Kalpana D, Omkumar KS, Kumar SS, Renganathan NG. A novel high power symmetric ZnO/carbon aerogel composite electrode for electrochemical supercapacitor. 2006;**52**:1309-1315. DOI: 10.1016/j.electacta.2006.07.032

[27] Mao L, Zhang K, On Chan HS, Wu J. Nanostructured MnO_2/graphene composites for supercapacitor electrodes: The effect of morphology, crystallinity and composition. Journal of Materials Chemistry. 2012;**22**(5):1845-1851. DOI: 10.1039/c1jm14503g

[28] Pu B et al. Flexible supercapacitors based on carbon nanomaterials. Journal of Materials Chemistry A. Materials for Energy and Sustainability. 2014;**3**(1): 22507-22541. DOI: 10.1016/j.mattod.2015.01.002

[29] Wang X, Dan W, Song X, Wei D, Zhao X, Zhang D. Review on carbon/polyaniline hybrids: Design and synthesis for supercapacitor. Molecules. 2019;**24**(12):2263. DOI: 10.3390/molecules24122263

[30] Choudhary N et al. Asymmetric supercapacitor electrodes and devices. Advanced Materials. 2017;**29**(21). DOI: 10.1002/adma.201605336

[31] Jeong GH, Baek S, Lee S, Kim SW. Metal oxide/graphene composites for supercapacitive electrode materials. Chemistry—An Asian Journal. 2016; **11**(7):949-964. DOI: 10.1002/asia.201501072

[32] Zeng Y, Yu M, Meng Y, Fang P, Lu X, Tong Y. Iron-based supercapacitor electrodes: Advances and challenges. Advanced Energy Materials. 2016;**6**(24): 1-17. DOI: 10.1002/aenm.201601053

[33] Yang S, Song X, Zhang P, Sun J, Gao L. Self-assembled α-Fe_2O_3 mesocrystals/graphene nanohybrid for enhanced electrochemical capacitors. Small. 2014; **10**(11):2270-2279. DOI: 10.1002/smll.201303922

[34] Li J, Chen D, Wu Q. α-Fe_2O_3 based carbon composite as pure negative electrode for application as supercapacitor. European Journal of Inorganic Chemistry. 2019;**2019**(10): 1301-1312. DOI: 10.1002/ejic.201900015

[35] Lorkit P, Panapoy M, Ksapabutr B. Iron oxide-based supercapacitor from ferratrane precursor via sol-gel-hydrothermal process. Energy Procedia. 2014;**56**(C):466-473. DOI: 10.1016/j.egypro.2014.07.180

[36] Shivakumara S, Penki TR, Munichandraiah N. High specific surface area α-Fe_2O_3 nanostructures as high performance electrode material for supercapacitors. Materials Letters. 2014; **131**:100-103. DOI: 10.1016/j.matlet.2014.05.160

[37] Shivakumara S, Penki TR, Munichandraiah N. Synthesis and characterization of porous flowerlike? Fe_2O_3 nanostructures for supercapacitor application. ECS Electrochemistry Letters. 2013;**2**(7):2013-2015. DOI: 10.1149/2.002307eel

[38] Abad ZAK, Nemati A, Khachatourian AM, Golmohammad M. Synthesis and characterization of rGO/Fe_2O nanocomposite as an efficient supercapacitor electrode material. Journal of Materials Science: Materials in Electronics. 2020;**31**(17):14998-15005. DOI: 10.1007/s10854-020-04062-7

[39] Li T et al. Facile electrochemical fabrication of porous Fe_2O_3 nanosheets for flexible asymmetric supercapacitors. Journal of Physical Chemistry C. 2017; **121**(35):18982-18991. DOI: 10.1021/acs.jpcc.7b04330

[40] Jiang X, Shi G, Wang G, Mishra P, Du J, Zhang Y. Fe_2O_3/hemp straw-based porous carbon composite for supercapacitor electrode materials. Ionics (Kiel). 2020;**26**(8):4039-4051. DOI: 10.1007/s11581-020-03547-z

[41] Chen L, Liu D, Yang P. Preparation of α-Fe_2O_3/rGO composites toward supercapacitor applications. RSC Advances. 2019;**9**(23):12793-12800. DOI: 10.1039/c9ra01928f

[42] Dong Y, Xing L, Chen K, Wu X. Porous α-Fe_2O_3@C nanowire arrays as flexible supercapacitors electrode materials with excellent electrochemical performances. Nanomaterials. 2018;**8**(7): 1-10. DOI: 10.3390/nano8070487

[43] Sarkar D, Pal S, Mandal S, Shukla A, Sarma DD. α-Fe_2O_3-based core-shell-nanorod–structured positiveand negative electrodes for a high-performance α-Fe_2O_3/C//α-Fe_2O_3/MnO x asymmetric supercapacitor. Journal of the Electrochemical Society. 2017;**164** (12):A2707-A2715. DOI: 10.1149/2.1711712jes

[44] Zhang M, Li X, Wang X, Li D, Zhao N. Three-dimensional core-branch α-Fe_2O_3@NiO/carbon cloth heterostructured electrodes for flexible supercapacitors. Frontiers in Chemistry. 2020;**7**(January):1-10. DOI: 10.3389/fchem.2019.00887

[45] Liu L, Wang J, Wang C, Wang G. Facile synthesis of graphitic carbon nitride/nanostructured α-Fe_2O_3 composites and their excellent electrochemical performance for supercapacitor and enzyme-free glucose detection applications. Applied Surface Science. 2016;**390**:303-310. DOI: 10.1016/j.apsusc.2016.08.093

[46] Zhong Y, Liu J, Lu Z, Xia H. Hierarchical FeS_2 nanosheet@Fe_2O_3 nanosphere heterostructure as promising electrode material for supercapacitors. Materials Letters. 2016;**166**:223-226. DOI: 10.1016/j.matlet.2015.12.092

[47] Zeng YX et al. Advanced Ti-doped Fe_2O_3@PEDOT core/shell anode for high-energy asymmetric supercapacitors. Advanced Energy Materials. 2015;**5**(12):1-7. DOI: 10.1002/aenm.201402176

[48] Gupta A et al. Nanostructured polyaniline/graphene/Fe_2O_3 composites hydrogel as a high-performance flexible supercapacitor electrode material. ACS Applied Energy Materials. 2020;**3**(7):6434-6446. DOI: 10.1021/acsaem.0c00684

[49] Guan C et al. Iron oxide-decorated carbon for supercapacitor anodes with ultrahigh energy density and outstanding cycling stability. ACS Nano. 2015;**9**(5):5198-5207. DOI: 10.1021/acsnano.5b00582

[50] Le K et al. Polypyrrole-coated Fe_2O_3 nanotubes constructed from nanoneedles as high-performance anodes for aqueous asymmetric supercapacitors. Dalton Transactions. 2020;**49**(28):9701-9709. DOI: 10.1039/d0dt01242d

[51] Zhao P et al. Facile hydrothermal fabrication of nitrogen-doped graphene/Fe_2O_3 composites as high performance electrode materials for supercapacitor. Journal of Alloys and Compounds. 2014;**604**:87-93. DOI: 10.1016/j.jallcom.2014.03.106

[52] Quan H, Cheng B, Xiao Y, Lei S. One-pot synthesis of α-Fe_2O_3 nanoplates-reduced graphene oxide composites for supercapacitor application. Chemical Engineering Journal. 2016;**286**:165-173. DOI: 10.1016/j.cej.2015.10.068

[53] Raza W, Ali F, Raza N, Luo Y, Kim K, Yang J. Recent advancements in supercapacitor technology nano energy recent advancements in supercapacitor technology. Nano Energy. 2018;**52**(August):441-473. DOI: 10.1016/j.nanoen.2018.08.013

[54] Abdel Maksoud MIA, Fahim RA, Shalan AE, Elkodous MA, Olojede SO, Osman AI, et al. Advanced materials and technologies for supercapacitors used in energy conversion and storage: A review. Environmental Chemistry Letters. 2021;**19**(1):375-439. DOI: 10.1007/s10311-020-01075-w

[55] Zhi M, Xiang C, Li J, Li M, Wu N. Nanostructured carbon–metal oxide composite electrodes for supercapacitors: A review. 2013:72-88. DOI: 10.1039/c2nr32040a

[56] Amirul M, Mohd A, Hawa N, Azman N, Kulandaivalu S, Sulaiman Y. Review of the use of transition-metal-oxide and conducting polymer-based fibres for high-performance supercapacitors.

Materials and Design. 2020;**186**:108199. DOI: 10.1016/j.matdes.2019.108199

[57] Huang M, Li F, Dong F, Zhang YX, Zhang LL. MnO_2-based nanostructures for high-performance supercapacitors. Journal of Materials Chemistry A. 2015;**3**(43):21380-21423. DOI: 10.1039/c5ta05523g

[58] Nithya VD, Sabari Arul N. Progress and development of Fe_3O_4 electrodes for supercapacitors. Journal of Materials Chemistry A. 2016;**4**(28):10767-10778. DOI: 10.1039/c6ta02582j

[59] Fan H, Niu R, Duan J, Liu W, Shen W. Fe_3O_4@carbon nanosheets for all-solid-state supercapacitor electrodes. ACS Applied Materials & Interfaces. 2016;**8**(30):19475-19483. DOI: 10.1021/acsami.6b05415

[60] Wang L et al. Three-dimensional macroporous carbon/Fe_3O_4-doped porous carbon nanorods for high-performance supercapacitor. ACS Sustainable Chemistry & Engineering. 2016;**4**(3):1531-1537. DOI: 10.1021/acssuschemeng.5b01474

[61] Kumar R, Singh RK, Vaz AR, Savu R, Moshkalev SA. Self-assembled and one-step synthesis of interconnected 3D network of Fe_3O_4/reduced graphene oxide nanosheets hybrid for high-performance supercapacitor electrode. ACS Applied Materials & Interfaces. 2017;**9**(10):8880-8890. DOI: 10.1021/acsami.6b14704

[62] Wu NL, Wang SY, Han CY, Wu DS, Shiue LR. Electrochemical capacitor of magnetite in aqueous electrolytes. Journal of Power Sources. 2003;**113**(1):173-178. DOI: 10.1016/S0378-7753(02)00482-2

[63] Wang S-Y, Ho K-C, Kuo S-L, Wu N-L. Investigation on capacitance mechanisms of Fe_3O_4 electrochemical capacitors. Journal of the Electrochemical Society. 2006;**153**(1):A75. DOI: 10.1149/1.2131820

[64] Liu J, Liu S, Zhuang S, Wang X, Tu F. Synthesis of carbon-coated Fe_3O_4 nanorods as electrode material for supercapacitor. Ionics (Kiel). 2013;**19**(9):1255-1261. DOI: 10.1007/s11581-013-0857-6

[65] Guan D et al. Hydrothermal synthesis of carbon nanotube/cubic Fe_3O_4 nanocomposite for enhanced performance supercapacitor electrode material. Materials Science & Engineering, B: Solid-State Materials for Advanced Technology. 2013;**178**(10):736-743. DOI: 10.1016/j.mseb.2013.03.010

[66] Lin TW, Dai CS, Hung KC. High energy density asymmetric supercapacitor based on NiOOH/Ni_3S_2/3D graphene and Fe_3O_4/graphene composite electrodes. Scientific Reports. 2014;**4**:1-10. DOI: 10.1038/srep07274

[67] Liu D et al. Ultrathin nanoporous Fe_3O_4-carbon nanosheets with enhanced supercapacitor performance. Journal of Materials Chemistry A. 2013;**1**(6):1952-1955. DOI: 10.1039/c2ta01035f

[68] Mitchell E et al. Facile synthesis and morphogenesis of superparamagnetic iron oxide nanoparticles for high-performance supercapacitor applications. New Journal of Chemistry. 2014;**38**(9):4344-4350. DOI: 10.1039/c4nj00741g

[69] Brousse T, Bélanger D. A hybrid Fe_3O_4-MnO_2 capacitor in mild aqueous electrolyte. Electrochemical and Solid-State Letters. 2003;**6**(11):4-9. DOI: 10.1149/1.1614451

[70] Qi T, Jiang J, Chen H, Wan H, Miao L, Zhang L. Synergistic effect of Fe_3O_4/reduced graphene oxide nanocomposites for supercapacitors with good cycling life. Electrochimica Acta. 2013;**114**:674-680. DOI: 10.1016/j.electacta.2013.10.068

[71] Du X, Wang C, Chen M, Jiao Y, Wang J. Electrochemical performances of nanoparticle Fe_3O_4/activated carbon supercapacitor using KOH electrolyte sol. Journal of Physical Chemistry C. 2009;**113**(6):2643-2646. DOI: 10.1021/jp8088269

[72] Wang L, Ji H, Wang S, Kong L, Jiang X, Yang G. Preparation of Fe_3O_4 with high specific surface area and improved capacitance as a supercapacitor. Nanoscale. 2013;**5**:3793-3799. DOI: 10.1039/C3NR00256J

[73] Meng W, Chen W, Zhao L, Yang H, Zhu M, Huang Y, et al. Porous Fe_3O_4/carbon composite electrode material prepared from metal-organic framework template and effect of temperature on its capacitance. Nano Energy. 2014;**8**:133-140. DOI: 10.1016/j.nanoen.2014.06.007

[74] Chen J, Huang K, Liu S. Hydrothermal preparation of octadecahedron Fe_3O_4 thin film for use in an electrochemical supercapacitor. Electrochimica Acta. 2009;**55**(1):1-5. DOI: 10.1016/j.electacta.2009.04.017

[75] Shi W et al. Achieving high specific charge capacitances in Fe_3O_4/reduced graphene oxide nanocomposites. Journal of Materials Chemistry. 2011;**21**(10):3422-3427. DOI: 10.1039/c0jm03175e

[76] Li X, Zhang L, He G. Fe_3O_4 doped double-shelled hollow carbon spheres with hierarchical pore network for durable high-performance supercapacitor. Carbon N. Y. 2016;**99**:514-522. DOI: 10.1016/j.carbon.2015.12.076

[77] Chaudhari NK, Chaudhari S, Yu J-S. Cube-like α-Fe_2O_3 supported on ordered multimodal porous carbon as high performance electrode material for supercapacitors. ChemSusChem. 2014;**7**(11):3102-3111. DOI: 10.1002/cssc.201402526

[78] Li Y, Jing X, Feng T, Yao Q, Xie J, Xia H. Fe_2O_3 Nanoneedles on ultrafine nickel nanotube arrays as efficient anode for high-performance asymmetric supercapacitors. Advanced Functional Materials. 2017;**27**(14):1606728. DOI: 10.1002/adfm.201606728

[79] Owusu K, Qu L, Li J, et al. Low-crystalline iron oxide hydroxide nanoparticle anode for high-performance supercapacitors. Nature Communications. 2017;**8**:14264. DOI: 10.1038/ncomms14264

[80] Mu J et al. Highly dispersed Fe_3O_4 nanosheets on one-dimensional carbon nanofibers: Synthesis, formation mechanism, and electrochemical performance as supercapacitor electrode materials. Nanoscale. 2011;**3**(12):5034-5040. DOI: 10.1039/c1nr10972c

Chapter 7

Green Energy Applications of Hematite (α-Fe_2O_3), Magnetite (Fe_3O_4), and Maghemite (γ-Fe_2O_3) Nanoparticles Based Hydroelectric Cell

Kuldeep Chand Verma and Navdeep Goyal

Abstract

Recently invented hydroelectric cell (HEC) is emerging as a better alternative for green electrical energy devices. HEC is fabricated as to generate electricity via splitting of water into H_3O^+ and OH^- ions without releasing any toxic product. In iron oxides, Hematite (α-Fe_2O_3), magnetite (Fe_3O_4) and maghemite (γ-Fe_2O_3) nanoparticles HEC are recently reported for their remarkable electrical response by splitting water molecules. Fe_3O_4 HEC 4.8 cm^2 surface size has delivered 50 mA short circuits current. Li ions into Fe_3O_4 stabilize electrical cell response to 44.91 mA with open-circuit voltage 0.68 V. Maghemite based HEC delivered a maximum short circuit current 19 mA with emf 0.85 V using water 200 μL. Maximum off-load output power 27.6 mW has been delivered by 4.84 cm^2 area hematite-HEC which is 3.52 times higher with 7.84 mW power as generated by Li-Mg ferrite HEC. Maximum electrical power 16.15 mW delivered by maghemite HEC is 0.58, 0.42 times lower than respective magnetite, hematite HECs. In more applicability of iron oxides, the multiferroic nanocomposites of $BaTiO_3$ with 85% $CoFe_2O_4$ has been shown maximum short circuit current 7.93 mA and 0.7 V emf by sprinkling few drops of water on HEC surface. $Li_{0.3}Ni_{0.4}Fe_{2.3}O_4$ and $Mg_{0.8}Li_{0.2}Fe_2O_4$ HECs also have some remarkable results for green energy generation.

Keywords: iron oxide nanoparticles, water splitting, green energy, oxygen vacancies

1. Introduction

World is fast becoming a global village due to the increasing daily requirement of energy by all population across the world while the earth in its form cannot change. Energy demand necessities for all societies to be services of energy to meet basic human needs like health, lighting, cooking, space comfort, mobility and communication, and serve as generative processes [1]. World's growing energy need, alongside increasing population led to the continual use of fossil fuel-based energy sources (coal, oil, and gas) which became problematic by creating several challenges. It includes depletion of fossil fuel reserves, greenhouse gas emissions and other environmental concerns, geopolitical and military conflicts, and the continual fuel price fluctuations. Recently, renewable energy sources are the most outstanding alternative and the only solution to the growing challenges of energy. In 2012, renewable energy sources supplied 22% total world energy generation which was not possible a decade ago. Renewable energy is an abundant source of energy and environment friendly as well [2]. Solar and wind energy for instance, are currently doing so well but their intermittency requires that energy storage or converting device system become more efficient and cost effective. Fuel cells are emerging energy converting devices which has low or no environmental effect but with the actual energy efficiency and energy density are lower than theoretical ones. Similarly dye synthesized solar cells are attractive energy conversion device have some limitations due to their stability. Renewables remained fastest growing source of energy in buildings, increasing 4.1% annually on average between 2009 and 2019. Use of renewable electricity for heat, that is, electric heat pumps, provided the second largest renewable energy contribution in recent years. Meantime, more than 256 gigawatts (GW) of renewable power capacity were added globally during recent year, surpassing the previous record by nearly 30.4% [1, 3, 4]. **Table 1** has shown present status of power capacity. Global population without access to electricity continued to shrink, although 771 million people (10% of the world's population) still lacked electricity access in 2019, nearly 75% of them in sub-Saharan Africa.

1.1 Global energy transition, 2050

Figure 1(a) provides a geographical breakdown of the renewable power generation capacity [5]. China accounts for over one third, followed by the United States, India,

Energy sources	Energy conversion and usage options	Power capacity 2020 (in GW)
Renewable power capacity (not including hydropower)	Power generation	1668
Renewable hydropower	Power generation	1170
Solar	Photovoltaic, thermal power generation, and water heaters	760
Wind power	Power generation, wind generators, windmills, and water pump	743
Bio-power	Heat and power generation, pyrolysis, gasification, and digestion	145
Geothermal power	Urban heating, power generation, hydrothermal, and hot dry rock	14.1
Solar thermal power	Solar home systems, solar dryers, and solar cookers	6.2
Ocean power	Numerous design, barrage, and tidal stream	0.5

Table 1.
Renewable energy sources, their use, and power capacity (in 2020) [1, 3].

and the European Union. Around 85% renewables in the power sectors with a large share from recurrent solar PV and wind is not possible without some strong combination of flexible dispatchable power, transmission interconnection, storage, smart grids, and demand-side management. Innovative technologies, operational practices, market designs, and business models are needed. Digital technologies open up new opportunities that yield new forms of flexibility such as aggregators that bundle services from small systems into marketable packages or consumer real-time price signals. In 2017,

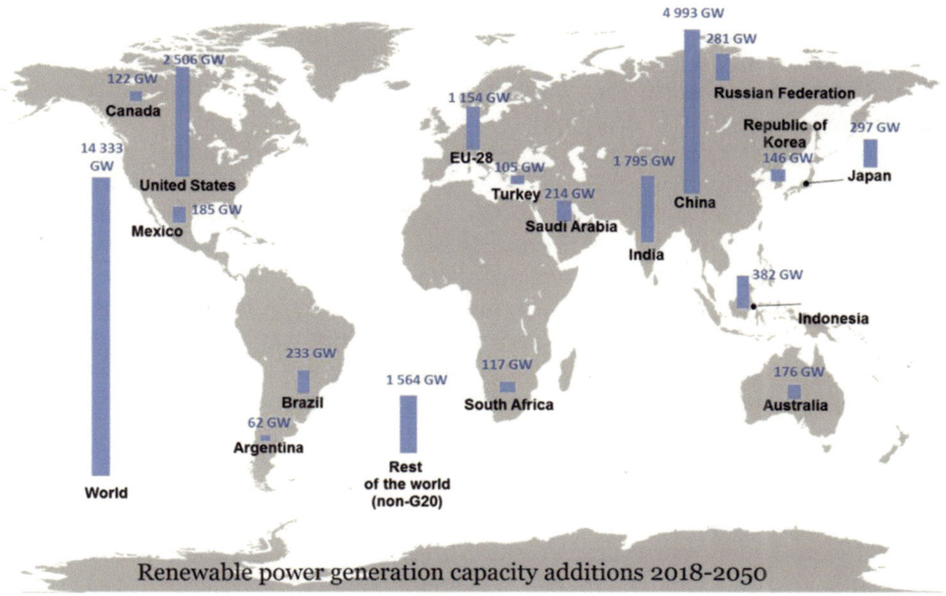

(a)

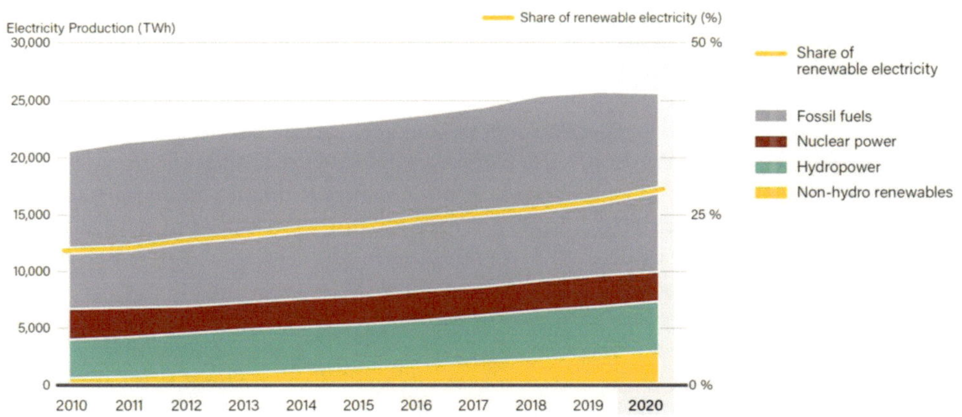

(b)

Figure 1.
(a) Geographical breakdown of renewable power generation capacity (2018–2050) [5]. (b) Global electricity production by source and share of renewables, 2010–2020 [3].

50 Hz grid operator in eastern Germany recorded an annual average 53.4% variable renewable energy. Growth of direct use of renewables in end use sectors (buildings, industry, and transport) would contribute 0.3% points annual renewables share growth, around a quarter of the total. Biomass alone would account for two-thirds of direct use of renewable energy in 2050. In primary energy terms annual bioenergy supply would roughly double from present levels to around 116 EJ in 2050.

1.2 Global electricity productions

In early 2020, global electricity demand dropped sharply in the wake of the COVID-19 pandemic. However, demand rebounded by year's end, resulting overall in a slight decline of around 2%, the first annual decline since the global economic crisis of 2008/2009. Production of electricity from renewables was favored under these low-demand circumstances due to its inherent low operating costs, as well as dispatch rules in many countries that prioritize renewable electricity. Renewables generated an estimated 29.0% of global electricity in 2020, up from 27.3% in 2019 (**Figure 1(b)**). Progress in renewable energy, and the decline in fossil fuels (coal), has been especially pronounced in certain countries and regions. Wind power, hydropower, solar power, and bioenergy became the EU-27's main source of electricity in 2020, growing from 30% of generation in 2015 to 38%. Electricity generation from these renewable sources grew 23% as production from coal power fell by half over this period. Similarly, in the United Kingdom, renewables grew to a 42% share of generation which become main source of electricity in 2020, beating out fossil gas and coal at 41%. In United States, renewable energy reached nearly 20% of net electricity generation, with solar and wind energy accounting for more than half of this in 2020. More than 19% of Australia's electricity came from wind and solar energy in 2020. In China, electricity from hydropower, solar energy, and wind energy provided more than 27% of production, up from around 26% in 2019. While variable renewables contributed more than 9% of global electricity in 2020, in some countries they met much higher shares of production, including in Denmark (63%), Uruguay (43%), Ireland (38%), Germany (33%), Greece (32%), Spain (28%), the United Kingdom (28%), Portugal (27%), and Australia (20%).

1.3 Renewable energy status in India

At present India is sixth largest country in the world in electricity generation with aggregate capacity 149 GWs out of which 25% hydro, 64% thermal, 3% nuclear, and about 8% renewable energy (small hydro, wind, cogeneration and biomass-based power generation, and solar) [6]. India's energy consumption has been increasing at a relatively fast rate because of economic development which grow at 8–9% per annum. Due to this India needs to attains a target of having 70% renewable energy use by 2050. Initiating with a very low base of renewable in 2000, the installed capacity of grid-connected renewable has reached 27.5 GW in 2013. **Figure 2** shows only 2.1% of renewable energy share to the Indian electricity grid. Government funded solar energy in India only accounted ~6.4 MW/year power as of 2005, 25.1 MW was added in 2010 and 468.3 MW in 2011. In view of these discussion about renewable energy sources, there is a need to develop more growth system for green energy from other sources like Hydroelectric cell (HEC) that recently investigated by Kotnala et al. on iron oxide nanoparticles [7–12].

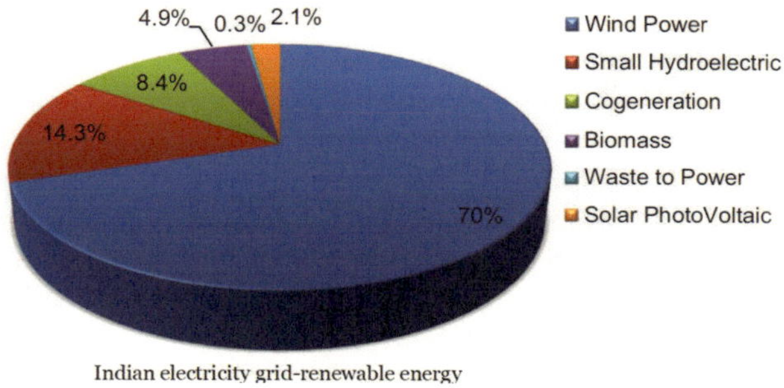

Figure 2.
Electricity grid-renewable energy share in India [6].

1.4 Hydroelectric cell as novel renewable energy source

Hydroelectric cell is newly invented as green energy source which offers many advantages over other renewable energy sources without using any solar/UV irradiation/electrolyte, while it uses little water for energy generation [7–13]. Most important thing towards green energy generation due to HEC is that the residues are non-toxic and its low-cost component raw materials. Electricity generation of HEC was first invented on $LiMgFe_2O_4$ due to water molecule dissociates on octahedrally under coordinated cations and oxygen defects [14]. Electrochemical redox reaction at respective Ag/Zn electrode with dissociated H_3O^+ and OH^- ions develop an emf 0.98 V. HEC is environmentally benign, low cost, easy manufacturing, facile in electricity generation with significantly useful by-products, and it is a potential candidate to replace existing portable green energy sources. HEC not required any other toxic chemicals unlike solar cell it can work in day or night and can run small scale devices like LED and fan. Two byproducts of HEC are hydrogen gas (highly pure 99.98% H_2 gas) and $Zn(OH)_2$ nanoparticles which has higher value into industrial commercial products and environment friendly.

1.5 Applications of hydroelectric cell

Green electricity production by hydroelectric cell has applications in geographically tough regions like rural areas, farms, forests, and mountains. It utilized for domestic and residential purposes in decentralized mode. Hydroelectric cell also be used as energy source in automotive industry. Using facile process, HEC can produce high-quality H_2 gas which can be stored for further use as a clean fuel. HEC acts as portable power generator for charging mobile phone, torch, video camera, laptop, etc. Furthermore, HECs can be used as power panels for stationary power generation. HEC produced high purity nanoparticles of zinc hydroxide and after thermally dissociation, ZnO nanoparticles are formed as the HEC byproduct for industrial applications.

1.6 Iron oxide based hydroelectric cell

Iron oxides have considerable potential into split water molecule and generate electricity through it. Defects in terms of oxygen vacancies of iron oxide nanoparticles make it suitable materials for HEC fabrication. Advantage of iron oxide nanoparticles is that the defects are easily formed during synthesis and suitable heating conditions that easily produce both Fe^{2+} and Fe^{3+} ions via oxygen vacancies [15–20]. Naturally occurring iron oxide like magnetite, maghemite, and hematite that easily splits water due to humid atmosphere [11]. Iron oxide nanoparticles have an amplified active surface suitable their reactivity towards polar water molecules. Water molecule interaction with iron oxide nanoparticles is highly dependent on coordination number of Fe ions along with surface composition. While dissociative adsorption of water is prominent on defective surface with coexisting Fe and O ions.

1.7 Iron oxides

Physical properties of iron oxides are given in **Table 2**. Iron oxides exist in eight forms in nature, with magnetite (Fe_3O_4), maghemite (γ-Fe_2O_3), and hematite (α-Fe_2O_3) which are probably most common due to their polymorphism involving temperature-induced phase transition [21]. Hematite is oldest known iron oxide that widespread in rocks and soils. Hematite has blood-red color that become black or gray in crystalline form. It extremely stable at ambient conditions, and often is the end product of the transformation of other iron oxides. Magnetite is also known as black iron oxide, ferrous ferrite, or Hercules stone and exhibits strongest magnetism. Maghemite occurs in soils as a weathering product of magnetite, or a product due to heating of other iron oxides.

1.7.1 Hematite (α-Fe_2O_3)

It is a most stable iron oxide under ambient conditions and can be used as a starting material for synthesis of magnetite and maghemite. Crystalline structure of hematite is shown in **Figure 3(a)**, where Fe^{3+} ions occupy two-thirds octahedral sites that confined by nearly ideal hexagonal close-packed O lattice. Oxygen ions are in a hexagonal close-packed arrangement, with Fe(III) ions occupying octahedral sites. Iron atom has a strong magnetic moment due to four unpaired electrons in 3d orbitals. In crystalline state, hematite is paramagnetic at temperatures above its Curie temperature 956 K. At room temperature, it is weakly ferromagnetic with phase transition 260 K (the Morin temperature, T_M) to an antiferromagnetic state [22]. Hematite Morin temperature decreases with decreasing particle size and tends to vanish below 8–20 nm. Recently investigated hematite HEC has remarkable results due to heterolytic dissociation of water into two surface hydroxyl groups is energetically more favorable [9].

1.7.2 Magnetite (Fe_3O_4)

As shown in **Figure 3(b)**, Fe_3O_4 has face centered cubic spinel structure (32 O^{2-} ions and close-packed along [111] direction) where all Fe^{2+} ions occupy half of the octahedral sites and Fe^{3+} are split evenly across the remaining octahedral and tetrahedral sites. Fe_3O_4 contains iron oxides into both divalent and trivalent states. Stoichiometric magnetite Fe^{II}/Fe^{III} = 1/2, and the divalent iron may partly or fully replaced by other divalent (Co, Mn, Zn, etc.) ions. Fe_3O_4 has lowest resistivity among iron oxides due to small bandgap 0.1 eV. Magnetite is ferrimagnetic at 300 K. Magnetite particles smaller than 6 nm are superparamagnetic at room temperature with coercivities ranging from 2.4 to 20 kA m^{-1} [23]. Dissociation of H_2O molecules at

Property	Fe oxides		
	Hematite	**Magnetite**	**Maghemite**
Molecular formula	α-Fe_2O_3	Fe_3O_4	γ-Fe_2O_3
Density	5.26 g cm^{-3}	5.18 g cm^{-3}	4.87 g cm^{-3}
Melting point	1350°C	1583–1597°C	—
Hardness	6.5	5.5	5
Magnetism	Weakly ferromagnetic or antiferromagnetic	Ferromagnetic	Ferrimagnetic
Curie temperature	956 K	850 K	820–986 K
M_S at 300 K	0.3 $Am^2\ kg^{-1}$	92–100 $Am^2\ kg^{-1}$	60–80 $Am^2\ kg^{-1}$
Crystallographic system	Rhombohedral, hexagonal	Cubic	Cubic or tetrahedral
Structural type	Corundum	Inverse spinel	Defect spinel
Space group	R3c (hexagonal)	Fd3m	$P4_332$ (cubic); $P4_12_12$ (tetragonal)
Lattice parameter (nm)	a = 0.5034, c = 1.375 (hexagonal), a_{Rh} = 0.5427, α = 55.3° (rhombohedral)	a = 0.8396	a = 0.83474 (cubic), a = 0.8347, c = 2.501 (tetragonal)
Band gap (E_g)	n-type; $E_g \sim 2.3$ eV	n- and p-type; $E_g \sim 0.1$ eV	n-type; $E_g \sim 2.0$ eV
Applications	Catalysts, pigments, gas sensors, and HEC	Solar energy conversion, context of imaging, drug delivery, environmental catalysis, recording devices, and HEC	Magnetic recording, high-frequency switch modes, electromagnetic absorbers, biomedical, and HEC

Table 2.
Physical and magnetic properties of iron oxides [8, 9, 11, 18, 19].

surface Fe sites and oxygen vacancies provides surface hydroxylation. On the hydroxylated surface, H_2O molecules get physisorbed and trap H_3O^+ ions inside mesopores to generate a high electric field, which further splits water molecules. This process results into increased ionic current in the cell. Li was doped into Fe^{3+} in Fe(II, III) oxide to create more oxygen vacancies to trap more electrons within oxygen vacancies [10].

1.7.3 Maghemite (γ-Fe_2O_3)

As shown in **Figure 3(c)**, maghemite has a spinel structure that is similar with magnetite irrespective to cation sublattice vacancies. Two-thirds sites are filled with Fe(III) ions arranged regularly, with two filled sites being followed by one vacant site. Oxygen anions give rise a cubic close-packed array while ferric ions distributed over tetrahedral sites (eight Fe ions per unit cell) and octahedral sites (remaining Fe ions and vacancies). Maghemite is ferrimagnetic at room temperature, unstable at high temperatures, and loses its susceptibility with time. Maghemite magnetic behavior is stabilized by doping of transition ions. Maghemite particles smaller than 10 nm are superparamagnetic at 300 K. Fe^{2+} ions oxidized into Fe^{3+} state in an ambient air by

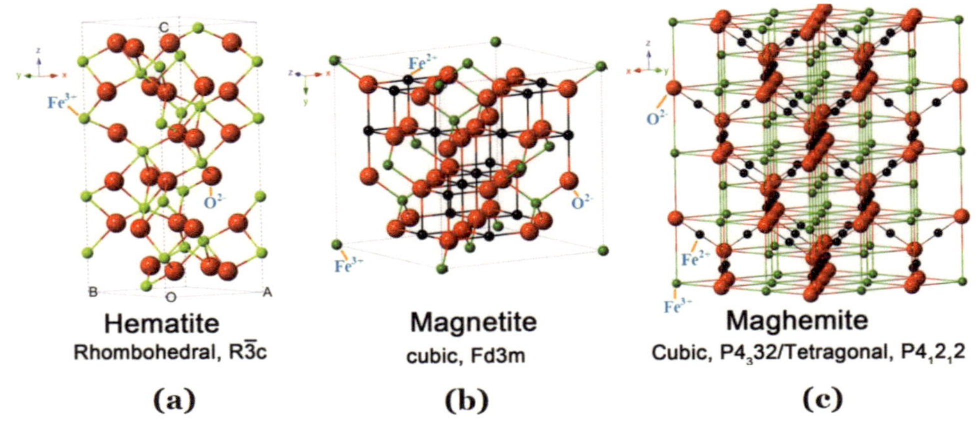

Figure 3.
Crystalline structure: (a) hematite, (b) magnetite, and (c) maghemite [19].

creating octahedral surface iron vacancies, $V^{Fe}{}_s$ for maintaining maghemite charge neutrality [11]. Non-stoichiometry in maghemite lattice is enhanced by creating oxygen vacancies with existing $V^{Fe}{}_s$ increases surface reactivity significantly.

1.8 Surface oxygen vacancies formations in iron oxides

Iron oxides include hematite, magnetite, and wustite (FeO) forms [24]. Magnetite Fe_2O_3 exhibits various polymorphs such as hematite α-Fe_2O_3 (rhombohedral), maghemite γ-Fe_2O_3 (cubic), β-Fe_2O_3 (cubic), and ε-Fe_2O_3 (orthorhombic), among which α-Fe_2O_3 is most thermodynamically stable phase. Both γ-Fe_2O_3 (space group: $P4_132$, $a = b = c = 0.8347$ nm) and Fe_3O_4 (space group: Fd3m, $a = b = c = 0.8394$ nm) share cubic structure with close-packed O atoms along <111> direction with vary Fe oxidation states. Applications for iron oxides intimately depend on their ability to redox (reduction and oxidation) cycle between +2 and + 3 oxidation states. Based on temperature-programmed reduction studies, two mechanisms have been proposed: a three-step mechanism, $Fe_2O_3 \rightarrow Fe_3O_4 \rightarrow FeO \rightarrow Fe$; and a two-step mechanism, $Fe_2O_3 \rightarrow Fe_3O_4 \rightarrow Fe$ [24]. In general, reduction of Fe_2O_3 is the hematite, does not occur directly metallic iron Fe. If the reduction temperature is lower than 570°C, reduction to Fe occurs stepwise from Fe_2O_3 to Fe_3O_4, called magnetite, and continues to Fe. Intermediate oxide, wustite $Fe_{1-x}O$, is not stable at temperatures lower than 570°C. For this, reduction occurs from Fe_2O_3 via Fe_3O_4 to $Fe_{1-x}O$ and continues afterward to Fe [25].

Even without addition from dopants or extrinsic defects, the properties of an oxide can be altered considerably if the material is made non-stoichiometric. Resulting, so-called intrinsic defects [vacancies at either O (V_O) or metal (V_M) site or interstitials of both types, O_i or M_i], alter number of electronic charge carriers. Defects formation change both geometric and electronic structure. When Fe_3O_4 is reduced, the non-stoichiometry is accommodated through Fe interstitials, and when it is oxidized, Fe vacancies created [26]. For hematite α-Fe_2O_3, Fe, and O vacancies are said to mediated water splitting process through localization of optically-derived charges. Adsorbed species and doping on the hematite surface, intrinsic vacancies can lead to formation of charge sites with consequent bending of the electronic valence and conduction bands. This increases the abundance of acceptor Fe vacancies [27]. Negreiros et al. [28] suggested hematite (α-Fe_2O_3) is a potential candidate for photo-electrochemical water splitting. Under oxygen rich conditions, the hematite, Fe_2O_3 (0001) terminated surface more stable [29]. On this surface termination, the isolated water molecule forms a heterolytically dissociated structure with OH^- group attached with surface Fe^{3+} ion and proton to surface O^{2-} ion. In contrast, in corundum, the vacancy site is filled with two electrons that repel OH^- ions. Here, the proton resulting from dissociated water forms a hydride ion (H^-). In the present chapter we have given a short review description on renewable energy sources in the form of hydroelectric cell based on iron oxides.

2. Synthesis of iron oxides hydroelectric cells

2.1 Co-precipitation technique

2.1.1 Magnetite HEC

Anhydrous $FeCl_3$ (32.44 g) and $FeCl_2{\cdot}2H_2O$ (16.22 g) (in molar ratio 1:2) were mixed in deionized water and the solution magnetically stirred at 60°C/1 h to get hydroxide precipitates by the addition of aqueous ammonia. Filtered precipitate was washed with deionized water and acetone to neutralize its pH. Black powder further vacuum-dried at 40°C/5 h. Dried powder pressed into 4.8 cm^2 pellet by a hydraulic press. Ag paste and Zn sheet also used as electrodes. Electrical contacts were made from the Ag and Zn electrodes to test the HEC performance [8, 9, 11].

2.1.2 Hematite HEC

Magnetite nanoparticles were synthesized by co-precipitation technique as discussed above. Filtered and washed precipitates with neutral pH are annealed at 500°C/2 h to obtain hematite nanoparticles. Crystalline hematite powder grinded and pressed into $2.5 \times 2.5 \times 0.1$ cm^3 pellet using hydraulic press. Pellet was further annealed at 600°C/2 h. Ag electrode in comb pattern was screen printed on one face of pellet and its second face was pasted with Zn anodic sheet of 0.3 mm thickness (**Figure 4(b)** and **(c)**) [30]. Measurement setup for HEC performance is shown in **Figure 4(a)**.

2.1.3 Maghemite HEC

Maghemite nanoparticles were synthesized by the oxidation of magnetite nanoparticles using a co-precipitation technique as discussed above. Filtered and washed precipitates annealed at 300°C/2 h. Oxidized brown powder was grinded and pressed into a $2.48 \times 2.48 \times 0.1$ cm^3 pellet using hydraulic press and further annealed at 350°C/2 h Ag paste and Zn sheet used as electrodes.

2.2 $Mg_{0.8}Li_{0.2}Fe_2O_4$ HEC using solid state reaction method

$Mg_{0.8}Li_{0.2}Fe_2O_4$ pellet was synthesized by solid state reaction method. High-purity precursors $MgCO_3$, Li_2CO_3, and Fe_2O_3 were taken in 0.8:0.1:1 molar ratio. Precursor powders mixed and grinded for 2 h in pestle and mortar and annealed at 850°C/10 h. Crystalline powder was grinded for 30 min and pressed into a $2.2 \times 2.2 \times 0.1$ cm^3 pellet and 4.6 cm diameter circular pellet of 0.1 cm thickness. Pellets were subjected

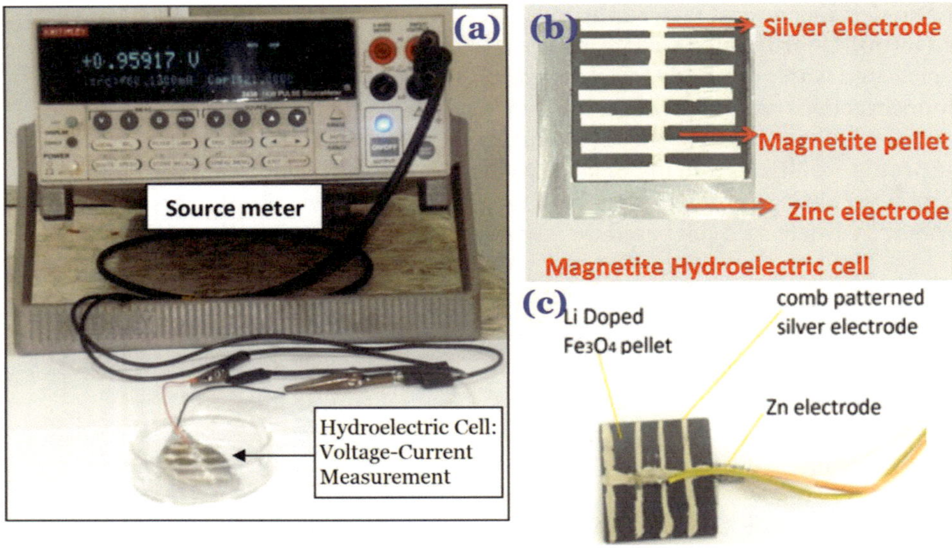

Figure 4.
(a) Experimental setup for V-I process in HEC using Keithley source meter (multiferroic $BaTiO_3/CoFe_2O_4$) [7, 30]. (b) Magnetite HEC with Ag comb pattern and Zn electrodes on the faces of pellet [8]. (c) Fabricated HEC in Li-doped Fe_3O_4 [10].

to sintering 1050°C/6 h. Comb-patterned Ag of 0.1 μm thickness and Zn sheet used as electrode of the pellet [12, 31].

2.3 $BaTiO_3/CoFe_2O_4$ HEC using sol-gel method

Ethanol and acetic acid are mixed in 75:25 ratio and $C_{16}H_{36}O_4Ti$ is added to it. A required amount of $Ba(CH_3COO)_2$, $CoCl_2{\cdot}6H_2O$, $FeCl_3$ are mixed in distilled water and then added to the precursor of $C_{16}H_{36}O_4Ti$ to get solution M. Solution M and PVA are mixed in 5:2 ratio and dried at 700°C/7 h. Pellet samples heated at 800°C/2 h. To fabricate this multiferroic HEC, the BTO-CFO crystalline powder was mixed with 3–5 drops of PVA solution (as binder) and then pressed into 4.5 cm^2 circular pellet of thickness 1 mm using hydraulic press with 5 bar pressure. BTO-CFO pellets finally sintered at 800°C/2 h. In order to increase porosity in pellet nanocomposite, more quantity of binder PVA was added and sintering temperature may reduce.

3. Results and discussion

3.1 Hematite HEC for generating electricity by water splitting

Hematite HEC delivered highest 30 mA current with an emf 0.92 V using ~500 μL deionized water. Rhombohedral structured hematite is confirmed with XRD pattern [9]. Values of lattice constant, a = 5.034 Å and c = 13.757 Å is matched well with corundum structure. Hematite average particle size is 20.76 nm as calculated with Scherrer relation. FESEM confirm large number of evenly distributed mesopores which might be due to coalescence and strong agglomeration of hematite nanoparticles with value of average pore size is 17 nm. Existence of characteristic A_{1g}

bands at 221 and 491 cm^{-1} and E_g bands at 239, 287, 401, and 605 cm^{-1}, respectively, ascribed with hematite structure using Raman spectroscopy. E_g bands at 401 and 605 cm^{-1} exists due to symmetric mode of O atoms related with cations in a plane perpendicular to crystallographic *c*-axis and Fe-O stretching vibrations, respectively [32]. FTIR spectra have investigated two IR absorption bands at 536 and 464 cm^{-1} are the characteristic of Fe-O stretching vibrations of crystalline hematite [33]. The stretching and bending vibrations of adsorbed water molecules with bands located around 3270 and 1624 cm^{-1} are observed [34]. Decreased intensity and slight Fe-O band shifting towards lower wavenumbers in wet hematite indicates change in Fe-O bond distance due to significant water adsorption. In XPS study, the high-resolution Fe 2p core-level spectra consisted of Fe 2p peak splitting into Fe $2p_{3/2}$ peak at 711.2 eV and Fe $2p_{1/2}$ peak at 724.6 eV is observed. Associated satellite peaks of Fe $2p_{3/2}$ and Fe $2p_{1/2}$ around 719.3 and 733.1 eV, respectively at ~8.1 and 8.5 eV high energy indicates presence of Fe^{3+} ions. O 1 s spectra leading to three components O_I, O_{II}, and O_{III} centered at 530.1, 531.3, and 532.6 eV, respectively is observed. Binding energy O_{II} peak around 531.3 eV is attributed with oxygen defects/vacancies. Broad PL emission band centered at ~564 nm (2.2 eV) corresponds to band edge transition. Non-radiative peaks at 656 nm (1.89 eV), 682 nm (1.82 eV), and 753 nm (1.65 eV) appeared from the transitions of trapped electrons in different defect states of oxygen vacancies on the surface of nanostructured α-Fe_2O_3.

3.1.1 TEM analysis

High-resolution TEM micrograph as given in **Figure 5(a)** depicted disoriented irregular hexagonal shaped nanoparticles with bright-dark fringes. Value of average grain size is 18.8 nm. Polycrystalline surface lattice fringes with 0.25 and 0.22 nm interplanar spacing correspond to (110) and (113) planes of hematite, respectively detected (**Figure 5(a′)**). Moire fringe pattern is observed in **Figure 5(a″)** which might be due to overlapping of strained crystallite fringes. Strain is relaxed due to structural defects, and the morphology of these defects is strongly dependent on growth conditions [35]. Planar surface defects and vacancy defects results into Moire fringes [36].

3.1.2 Microstructural analysis

Specific surface area 72.2 m^2 g^{-1} has been calculated by adsorption isotherm using multipoint BET method (**Figure 5(b)**). Average pore width 16.8 nm is obtained from pore size distribution curve using Barrett-Joyner-Halenda (BJH) method. Cumulative pore volume 0.289 cc g^{-1} is obtained with pore diameter less than 39 nm indicating mesoporous structure. The value of porosity is 27.4% which has been calculated by volumetric ratio method.

3.1.3 Voltage-current conduction process

Figure 5(c) is the voltage-current (V-I) polarization that shows 30 mA short circuit current and an open cell potential 0.92 V of wet hematite HEC of size 4.84 cm^2. The highly electronegative surface Fe^{3+} cations act as Lewis acid attracting water molecules strongly due to interaction of oxygen lone pair electron with octahedrally coordinated surface Fe ions having empty 3d orbitals [37]. This is the process of heterolytic dissociation of water molecule occurs with OH^- ion bonded on surface Fe^{3+} ion and dissociated H^+ ion bonded nearby surface O^{2-} ion [38, 39]. Two OH^- groups

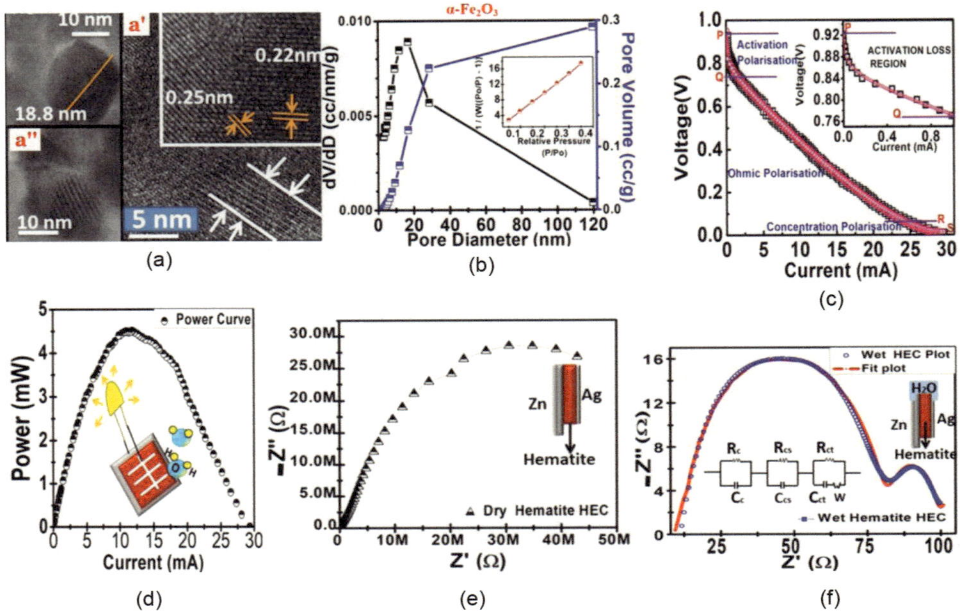

Figure 5.
(a) HRTEM image showing disoriented hexagonal hematite nanoparticle, (a') surface lattice fringe pattern (inset), unequal d spacing (white arrows). (a") Surface defect Moire fringe pattern, (b) BJH cumulative pore volume variation with pore size (inset is the active surface area of mesoporous), (c) V-I polarization curve of wet hematite HEC (inset is the low current activation polarization region), (d) respective electric power loss curve (schematic hematite HEC). Nyquist plot of (e) dry and (f) wet hematite HEC [9].

formed due to initial chemidissociation of single water molecule. The physisorption of water molecule occurs via hydrogen bonding on chemidissociated surface OH group layer and therefore, the physisorbed layers form H_3O^+ ions by H^+/H_2O ion due to hopping mechanism [40]. Surface defects due to oxygen vacancies significantly enhance the reactivity of hematite surface. In **Figure 5(c)**, the internal losses in HEC lead reduction of open cell voltage (E_{oc}) into 0.92 V. Since there is formation of different irreversible polarization losses (PQ, QR, and RS region) in HEC including activation polarization, ohmic polarization, and mass transport polarization which induces voltage drop with increment in operating current [41]. Therefore, the output voltage of HEC may describe by following Eq. (1):

$$V_{out} = E_{open\ cell} - V_{activation} - V_{ohmic} - V_{mass\ concentration} \tag{1}$$

where V_{out} is the output overvoltage and E_{oc} open cell voltage. V_{act}, V_{oh}, and V_{conc} are activation, ohmic, and mass transport overvoltage loss, respectively. To overcome the energy barrier of electrochemical reaction occurring at Ag/Zn electrodes due to initial surface chemidissociated ions, the activation overvoltage (region PQ) might be required which is 0.11 V at low current density [9]. Due to larger number of chemidissociated surface hydroxyl groups occurring at hematite surface added physisorb water molecules via hydrogen bonding. Concurrently high electric potential is generated in mesopores of hematite due to trapped H_3O^+ ions which responsible into spontaneous dissociation of physisorbed water molecules into H_3O^+ and OH^- ions. In the intermediate current region

BC, the linear voltage drop with increment in operating current is observed. This might be due to decrement in hematite cell resistance by spontaneous water molecule dissociation and charge/ion transport through crystalline boundaries. However, the voltage drop occurs due to overcoming loss due to flow of dissociated H_3O^+/OH^- ions through mesoporous electrode surface. Sudden voltage degradation observed at high current density region (CD) which is the effect of concentrated polarization overvoltage. In this case, the charge/ions transported in high concentration at electrode surface are quickly taken away by active Zn/Ag electrodes to flow high current. Maximum hematite HEC current 30 mA recorded which result by polarization losses fallow lower rate of charge/ion transport phenomenon. Therefore, the output current might be affected with mass transport overvoltage loss [42]. **Figure 5(d)** showed electrical power ($P_{out} = V \times I$) due to presence of polarization losses which is 4.5 mW with maximum off-load output power ($P_{max} = E_{oc} \times I_{sc}$) is 33.94 mW.

3.1.4 Ionic conduction by impedance spectroscopy

Dynamics of charge/ion transport from hematite surface-electrode interface in wet HEC is determined by Nyquist plot equivalent circuit as given in **Figure 5(e)** and **(f)** [inset of **Figure 5(f)** is the equivalent circuit]. Values of impedances in dry and wet state hematite HEC is given in **Table 3**. A single semicircular resistive arc in dry hematite HEC is observed with impedance in the order ~10^7 Ω which might be the effect of large number of defects and high porosity in the pellet sample. With few microliter of Millipore deionized water onto dry hematite HEC, the overall impedance is decreased 100 Ω. Equivalent circuit consists of R_c (resistance of crystallite); R_{cb} (resistance of crystallite boundary); R_{ct} (resistance by charge transport at the interface between electrode-hematite surface); W (Warburg element contribution by diffusion of ions at electrode surface); and C stand respective capacitance [43]. Fitted parameters from the equivalent circuit modeling for wet HEC are R_c (Ω) = 43, R_{sc} (Ω) = 23.7, R_{el} (Ω) = 19, C_c (F) = 5×10^{-9}, $n_1 = 0.84$, C_{sc} (F) = 4.5×10^{-7}, $n_2 = 0.68$, C_{el} (F) = 0.66, and $A_w = 32$. The higher frequency depressed semicircle is the bulk crystallite contribution where the water molecules chemidissociate at unsaturated surface cations. The middle frequency depressed semicircle is the effect of nanocrystallites boundaries contribution which has comparatively higher defect concentration to dissociate more water molecules. The single semicircle observed at lower frequency region corresponds to charge/ion transport phenomenon by ionic diffusion at crystallite surface/electrode interface.

3.2 Magnetite nanoparticles HEC

Typical magnetite HEC of 4.8 cm^2 area delivers 50 mA peak current with a maximum output power 38.5 mW. An electromotive force (emf) of 0.77 V is generated due to a redox reaction at respective electrodes. **Figure 6** shows the HEC results of magnetite [8] where the water molecule chemidissociates on Fe surface cations and oxygen vacancies due to physisorbed water molecule dissociation results into charges trapped inside mesopores. Magnetite crystalline structure by XRD pattern gives diffraction peaks at 30.4°, 35.43°, 37.09, 43.08°, 53.43°, 56.94°, and 62.57° correspond to (220), (311), (222), (400), (422), (511), and (440) lattice planes of spinel phase, respectively. Interplanar spacing of most intense Fe_3O_4 (311) peak is calculated to be 2.53 Å with the value of lattice constant 8.38 Å. Crystallite size using the Scherrer relation ~12 nm.

x	Synthesis method	HEC area size (cm^2)	Crystalline phase	D (nm)	Pore size/p (%)	I_{sc} (mA)	V_{oc} (V)	R_D (Ω)	R_W (Ω)
Fe_3O_4 [8]	Co-precipitation	4.8	Cubic spinel	11	10 nm/46	50	0.77	1×10^5	80
Li:Fe_3O_4 [10]	Co-precipitation	4.08	Cubic spinel	400	4 nm	44.91	0.68	1×10^3	71
γ-Fe_2O_3 [11]	Co-precipitation	4.84	Cubic spinel	9–17	10.78 nm	19	0.85	1×10^6	1000
α-Fe_2O_3 [9]	Co-precipitation	4.84	Rhombohedral	18	39 nm/27.4	30	0.92	1×10^7	100
$Mg_{0.8}Li_{0.2}Fe_2O_4$ [12]	Solid state reaction	4.8	Cubic spinel	400	30	8	0.98	1×10^8	100
$Li_{0.3}Ni_{0.4}Fe_{2.3}O_4$ [30]	Solid state reaction	4	Cubic spinel	47	2.5 nm	15.3	0.90	5.3×10^6	15.5
$Mg_{0.7}Li_{0.3}Fe_2O_4$ [31]	Solid state reaction	4.9	Inverse spinel	1000	—	17	0.859	—	—
$BaTiO_3/CoFe_2O_4$ [7]	Sol-gel using PVA	4.5	Tetragonal/Cubic spinel	70	2.3–10.8	1.2–7.9	0.7–0.97	0.5×10^6	5

Table 3.
Iron oxides HECs: Composition (x), synthesis method, cell area size (cm^2), crystalline phase, particles size (D), pore size/porosity (p), short circuit current (I_{sc}), open circuit voltage (V_{oc}), dry cell resistance, R_D (bulk), and wet cell resistance, R_W (bulk).

3.2.1 Defects/vacancies study by Raman analysis

As Raman spectra shown in **Figure 6(a)**, the five active bands at 200 (E_g), 317 (T_{2g}), 524 (T_{2g}), 533 (T_{2g}), and 668 (A_{1g}) cm^{-1} confirm the formation of spinel magnetite [44, 45]. Since oxidation of magnetite to maghemite is highly sensitive with laser power, therefore the maghemite peaks at 352 (T_{2g}) and 709 (A_{1g}) cm^{-1} may due to oxidation of magnetite. Raman spectral peaks of magnetite are shifted by 3–7 cm^{-1} towards higher frequency which may attribute by quantum confinement effect.

3.2.2 Structural and defects study by FTIR analysis

Figure 6(b) is the functional group study by FTIR spectroscopy which shown two distinct bands at 570 and 446 cm^{-1} due to intrinsic stretching vibration mode associated with Fe-O absorption bonds, respectively tetrahedral and octahedral sites of Fe_3O_4 spinel structure [46].

3.2.3 Nanopores and particles size by FESEM pattern

FESEM image of magnetite nanoparticles given in **Figure 6(c)** shows aggregation type product with nanopores formation. Elemental identification by EDX spectra is also given in the inset of **Figure 6(c)**. Presence of Fe and O peaks with 71/29 wt% and 39/61 atomic% confirms the magnetite stoichiometry. Fe_3O_4 average grain size is 11 nm with mesopores size <50 nm, along with some macropores of size >50 nm is

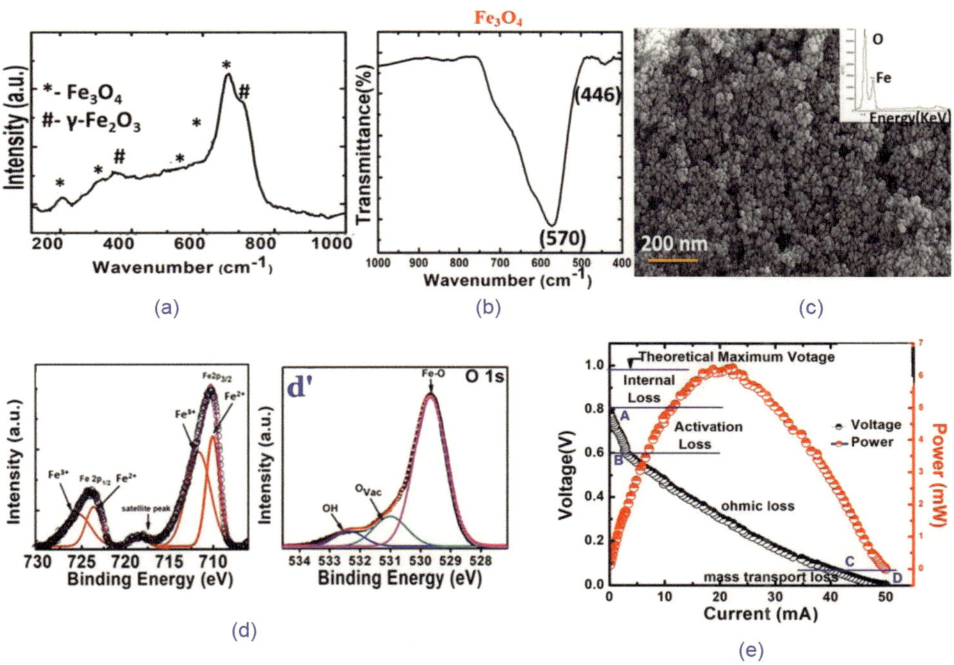

Figure 6.
(a) Raman spectra of magnetite Fe_3O_4 nanoparticles measured at room temperature. (b) FTIR spectroscopy of magnetite showing Fe-O bonds. (c) FESEM image showing porous grains (inset is the EDX spectrum). XPS spectra of Fe_3O_4: (d) Fe 2p and (d') corelevel O 1 s. (e) V-I characteristics and output power in wet deionized water condition [8].

detected. Nanopore size and specific surface area of present magnetite nanoparticles is determined with BET nitrogen adsorption-desorption isotherms [8]. Specific surface area of Fe_3O_4 is found to be 89.78 m^2g^{-1} and the average pore diameter is 9.81 nm. Cumulative pore volume with pore diameter < 190 nm is 0.335 cm^3g^{-1}. Using volumetric ratio analysis, the total porosity of Fe_3O_4 nanoparticles is estimated to be 46%.

3.2.4 Iron and oxygen vacancies confirmation with XPS analysis

Magnetite valence states of Fe and O are investigated with X-ray photoelectron spectroscopy as given in **Figure 6(d)** and **(d′)**. Satellite peak at 718.5 eV is attributed with Fe^{3+} ion indicates partial oxidation of magnetite into γ-Fe_2O_3 which might be possible when Fe^{2+} is subjected to air oxidation into Fe^{3+} state [47]. Mean relative area under the curve of two Fe 2p peaks confirms attributed to Fe^{2+} and Fe^{3+} ionic states which used to calculate the stoichiometric ratio of Fe^{2+}/Fe^{3+} ions, that is, 31.5:68.5 close to 33:67 (as standard stoichiometric ratio of Fe^{2+}/Fe^{3+} in Fe_3O_4). **Figure 6(d′)** is the O 1 s XPS spectra which clearly shown peaks at 529.7, 531, and 532.3 eV, respectively of Fe-O, oxygen vacancies and OH ions. A major peak around 529.7 eV with an integrated peak area of 77% is attributed with lattice oxygen of Fe_3O_4 [47]. The observed surface oxygen defects in terms of oxygen vacancies and hydroxyl species act as donor centers for water molecule dissociation.

3.2.5 Magnetite HEC V-I performance

Figure 6(e) shows the electrical results of magnetite HEC that partially dipped in deionized water and generates 50 mA short-circuit current and 0.77 V open-cell voltage. This observed electrical response has a typical polarization region [48]. Decrease in voltage with increasing current is attributed with kinetics of electrochemical cell having three prominent loss regions. At point A, the open-circuit voltage of 0.77 V is lower than the redox potential of Ag and Zn electrodes. Decrease in voltage may attribute due to internal anodic behavior of magnetite owing to octahedral Fe^{2+}-Fe^{3+} ionic oxidation. Activation loss (region AB) is dominant at lower current which results from delay in initiation of water dissociation process in mesoporous Fe_3O_4 surface and collection of dissociated ions by the electrodes. This activation energy is the energy barrier which is required to overcome electrochemical reaction that happening on mesoporous magnetite surface and electrode. Voltage drop in intermediate current density (region BC) is linear which is ohmic losses provides resistance to flow of ions through porous structure. Current density region CD has a sharp voltage drop which is observed due to crowding of electrode surface with high concentration of ions (mass transport loss) [42]. Output power has maximum value 38.5 mW is observed.

3.2.6 Mechanism for current conduction in magnetite HEC

Since the unsaturated Fe^{2+} and Fe^{3+} surface cations and oxygen vacancies in magnetite help to chemidissociation of water molecules. The high electron spin density makes more dangling bonds with trapped electrons as active sites for water molecule dissociation. The dissociated H_3O^+ ions get trapped inside Fe_3O_4 mesopores to generate high electric field. The trapped electrons pull nearer H_2O molecule and consequently, the unsaturated surface metal cations immediately take OH^- ion out of it. When the OH^- ion is taken away, the H_3O^+ ion remains, and these H_3O^+ ions are trapped within nanopores of HEC. It leads to generate very high electric field inside pores. Hydronium ions, H_3O^+ hops towards Ag electrode and hydroxide OH^- ions diffuse towards Zn electrode through surface vacancies. Redox reaction occurs between both electrodes leads to voltage and current generation in HEC due to Grotthuss process. Due to presence of nanosized pores on Fe_3O_4 surface, the H_2O molecules dissociates with high electrostatic potential by trapped H_3O^+ ions inside nanopores. Zn gets oxidized to produce $Zn(OH)_2$, and H_3O^+ ions are reduced at Ag cathode to the evolution of H_2 gas by capturing electrons from the Zn anode, generating emf in the cell due to reactions:

$$\textit{At magnetite surface}: 4H_2O \rightarrow 2H_3O^+ + 2OH^- \tag{2}$$

$$\textit{At Anode}: Zn + 2OH^- \rightarrow Zn(OH)_2 + 2e^- \tag{3}$$

$$\textit{At Cathode}: 2H_3O^+ + 2e^- \rightarrow H_2(g)\uparrow + 2H_2O \tag{4}$$

In this process of electric current generation, $Zn(OH)_2$ deposited at Zn anode. Hydrogen H_2 gas evolution at the Ag anode. Resistance of dry magnetite cell is found to be 10^5 Ω, which is reduced to 80 Ω after dipping the cell with deionized water.

3.3 Green energy enhancement with Li ions in magnetite HEC

The 4.08 cm^2 size HEC of Li-doped Fe_3O_4 ($Li_{0.4}Fe_{2.6}O_4$) delivers short circuit current, emf, and off-load output power is 44.91 mA, 0.68 V, 30.80 mW, respectively [10]. XRD pattern revealed spinel cubic structure with lattice constant 8.35 Å is measured. Using Debye-Scherrer relation, the value of crystallite size from most intense (311) peak is 8 nm. Porous microstructure using BET technique in FESEM analyzed the specific surface area to be 45 $m^2\ g^{-1}$, cumulative pore volume 0.04683 $cm^3\ g^{-1}$ and pore size 4 nm at a relative pressure 0.99663 (P/Po) are measured. Li ions in Fe_3O_4 have created more oxygen vacancies in the spinel lattice to enhance the capability for water dissociation. XPS analysis indicates the existence of Fe into +2 and + 3 oxidation states of Fe_3O_4 and monovalent Li^{1+} at the divalent Fe^{2+} site because to the concentration oxygen vacancy is enhanced. Oxygen vacancies accelerate the process of water splitting to produce larger current being generated by Li doping as compared with pure Fe_3O_4 HEC.

3.3.1 HEC V-I response

In **Figure 7(a)** the observed V-I mechanism is described with four types of different control segments [49]. Segment OP is the internal loss region where voltage 0.98 V as theoretical maximum exists due to internal losses in the cells. Segment PQ at low current density region is the activation loss where the voltage needed to overcome electrochemical reactions between Li:Fe_3O_4 surface cations and electrodes. Intermediate segment QR is the voltage reduction by cell ohmic losses which is mostly responsible to provide hindrance to the flow of ions via porous network. At high current density segment RS, a sudden decline in voltage is found is attributed due to highly reactive electrodes that do not get enough ions to react. This is also called concentration loss or mass transport loss [50]. On-load peak power shown by polarization plot is 5.39 mW.

3.3.2 Impedance measurement and Nyquist plots

The electrostatic impedance spectroscopy is an important tool to determine the dynamic behavior of a cell/battery [51]. A Nyquist Cole-Cole for Li-doped Fe_3O_4 HEC is shown in **Figure 7(b)** for both wet and dry states. It helps us to study the diffusion of ions due to water molecules splitting. From the equivalent circuit [inset of **Figure 7(b)**], the resistance R_1 corresponds to the first semicircle in the high frequency region and resistance R_2 from second semicircle in the middle frequency. Constant phase element Q is used for a capacitor to compensate non-ideal behavior of the electrode due to surface roughness or porosity and the Warburg element, W identify the diffusion of ions. Li ions into Fe_3O_4 HEC exhibits impedance in its dry state in the order of 10^3 Ω which reduces into 71 Ω in its wet state. The HEC in moist state displays a tail at a low frequency which confirms the ionic diffusion of H_3O^+.

3.4 HEC performance in maghemite

Due to non-stoichiometric behavior of maghemite, the water dissociation is highly performed to increase cell capacity [11]. The surface active sites oxygen vacancies, $V^O{}_s$ in magnetite and hematite HECs stimulated water dissociation to generate 50 and 30 mA current, respectively which is higher with maghemite HEC 19 mA. XRD pattern results into cubic spinel maghemite with lattice parameter, a = 8.35 Å. Vibrational Raman broad band are measured around 360 (T_{2g}), 500 (E_g), and 700 cm^{-1} (A_{1g}) which showed characteristic peaks of maghemite [11, 45]. The broad A_{1g} mode attributed due to vibration of O atoms along Fe-O bonds. The $V^{Fe}{}_s$ related with local surroundings in lattice resulting into different Fe-O distance near vacancy defects [52]. Band at 667 cm^{-1} is assigned to FeO_4 tetrahedra vibration of non-defective spinel while the band at 719 cm^{-1} exists due to Fe-O vibrations adjacent to $V^{Fe}{}_s$. From FTIR study, the strong IR band at ~570 cm^{-1} exists due to Fe-O vibrations which eventually broadens and splits into newer bands due to oxidation process. A strong IR band at 638 cm^{-1} along with a shoulder near 538 cm^{-1} is attributed with shifting and splitting of Fe-O stretching vibrational mode at (ν_1) ~570 cm^{-1} [53]. However, a weak absorption at ~440 cm^{-1} is assigned to shifting of octahedral Fe-O vibrational mode ν_2 [45]. The HRTEM study has revealed average grain size 13.1 nm nanopore size 11.6 nm.

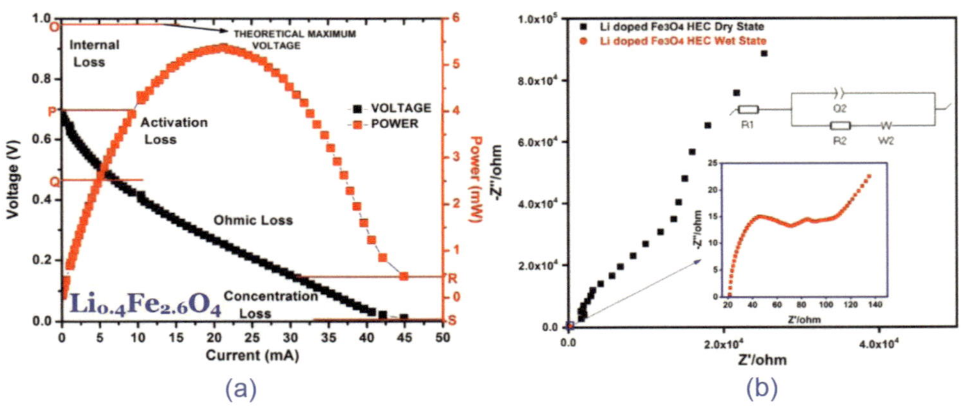

Figure 7.
(a) V-I characteristics in Li-doped Fe_3O_4 ($Li_{0.4}Fe_{2.6}O_4$) HEC in deionized water under wet condition [10]. (b) Corresponding Nyquist plot in both dry and wet states.

Using BET method, the value of effective surface area 90.37 $m^2\ g^{-1}$, and an average pore size 10.78 nm, pore volume 0.33 cc g^{-1} and maximum pore size 120 nm have been analyzed. XPS peaks obtained ~710.7 eV, 717.7 eV and 724.3 eV, 733 eV corresponded to respective Fe $2p_{3/2}$ and Fe $2p_{1/2}$ peaks, respectively, of maghemite [11, 54].

3.4.1 Lattice defects formation with photoluminescence

Figure 8(a) shows the photoluminescence (PL) emission peaks observed from nanostructured maghemite. The PL peak at ~550 nm (2.25 eV) is assigned with radiative recombination of electrons due to crystal field splitting energy levels in maghemite octahedral sites [55]. PL peak at 615 nm confirm the transition from tetrahedral crystal field splitting energy respect to O (2p). A hump at ~655 nm and strong PL peak at ~830 nm is attributed with radiative recombination of electron trap levels, respectively with tetrahedral and octahedral sites with O (2p) band which resulted by $V^O{}_s$. $V^{Fe}{}_s$ corresponds with octahedral sites acts as acceptor type surface defects may responsible with an additional peak at ~740 nm.

3.4.2 Charge transport kinetics with dielectric measurement

Since low conductivity of crystallite boundaries in maghemite arises due to presence of $V^{Fe}{}_s$ disrupting Fe^{2+}/Fe^{3+} hopping path breaking its electronic conduction [11]. As shown in **Figure 8(b)**, the magnetite pellet is found to be more conducting than maghemite even in its wet state. A sharp increment in capacitance is observed from dry to wet state (varying capacitance from 10 pF to 3 μF in maghemite compared with 100 nF to 2.8 μF in magnetite) along the electrode region. Comparatively higher capacitance (10^6 order) in maghemite inferred to the formation of larger doubly and triply hydrogen bounded physisorbed water molecules at $V^{Fe}{}_s$ than singly bound water molecules in magnetite. The V-I characteristics of maghemite HEC are given [11]. With the addition of 200 μL deionized water on 4.84 cm^2 area maghemite HEC generates 19 mA short circuit current with an emf 0.85 V (**Table 3**). In comparison to hematite and magnetite HECs, the less output electricity generated by maghemite

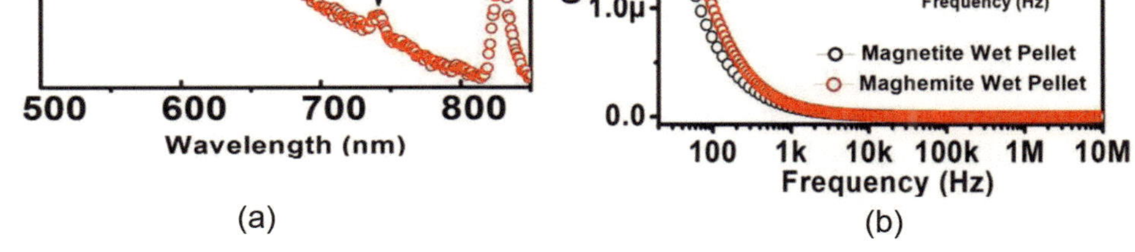

Figure 8.
(a) Photoluminescence spectra of maghemite γ-Fe_2O_3. (b) Variation of capacitance with frequency in its wet state (inset is the maghemite pellet with corresponding capacitive plot in its dry state) [11].

HEC may attributed by presence of acceptor type $V^{Fe}{}_s$ with reduced V^O concentration. The molecular H_2O adsorption on V^{Fe} sites is responsible to reduced chemi-dissociation rate of water in maghemite [56].

3.5 Progress on Li substituted $MgFe_2O_4$ HEC

The $Mg_{0.8}Li_{0.2}Fe_2O_4$ HEC cell of size 17 cm^2 is able to generate short circuit current 82 mA and 920 mV emf with maximum output power 74 mW [12]. This current conduction process occurs due to dissociated H_3O^+ and OH^- ions are transport through surface and capillary diffusion in porous ferrite towards Zn and Ag electrodes. **Figure 9(a)** is the mechanism for current conduction process in $Mg_{0.8}Li_{0.2}Fe_2O_4$ HEC [12]. Li^{1+} ions in $MgFe_2O_4$ create oxygen vacancy which acts as dangling/unsaturated bond to produce trapped electrons. Electric field developed inside the pore is physisorbed water molecule spontaneously. Further, voltage is generated by oxidation reaction occurring at Zn electrode and reduction of H_3O^+ occurring at Ag electrode due to Eqs. (2)–(4). Generated voltage helps to transport H_3O^+ and OH^- ions towards respective electrodes. Surface lattice fringe widths 0.25 and 0.16 nm correspond to (311) and (511) lattice planes, respectively for $Mg_{0.8}Li_{0.2}Fe_2O_4$ measured by HRTEM analysis [12]. Using BET method, the specific surface area of $Mg_{0.8}Li_{0.2}Fe_2O_4$ pellet is determined to be 165 $m^2\ g^{-1}$. Total pore volume for pores smaller than 455 nm diameter is obtained to be 0.74 cc g^{-1} along with 30% total porosity. $Mg_{0.8}Li_{0.2}Fe_2O_4$ HEC exhibits high reactance in its dry state (order of $10^8\ \Omega$). When HEC is partially dipped in deionized water, the reactance of cell pellet is decreased to ~100 Ω.

3.6 Progress on $Li_{0.3}Ni_{0.4}Fe_{2.3}O_4$ HEC

The Ni substituted lithium ferrite, $Li_{0.3}Ni_{0.4}Fe_{2.3}O_4$ (LNFO) HEC generate green electricity has been reported [30]. The Ni substituted at octahedral site of Li ferrite ($Li_{0.5}Fe_{2.5}O_4$) because the reasoning behind Ni^{+2} (0.72 Å) substitution at Li^{+1} (0.76 Å) and $Fe^{+3/2}$ (0.63/0.77 Å) site is the ionic radius and lower/higher valances. Due to occurrence of valance and ionic radii mismatch, the resulting lattice strain produces oxygen defects [7]. XRD pattern revealed cubic spinel phase with the value of lattice constant 8.319 Å and average particles size 47.8 nm is measured. From FESEM, the average grain size from pellet sample is 475 nm. From BET analysis, the value of pore size is 2.5 nm. FTIR has revealed the higher frequency absorption band ν_1 at 606 and 605 cm^{-1} is attributed due to stretching vibrations of tetrahedral sites. Lower frequency absorption band ν_3 found at 412 and 411 cm^{-1} is the stretching vibrations of octahedral sites [57, 58]. Middle frequency absorption band ν_2 at 479 cm^{-1} confirm the availability of Li-O complexes at octahedral sites. Absorption band at 3425 cm^{-1} corresponds to stretching vibrations of the surface adsorbed H_2O molecules.
PL emission observed peak at 472 nm corresponds to blue emission and 613 nm (orange emission) confirms to the presence of defects states within forbidden energy gap. PL emission peak at lower wavelength [around ~611 nm (2.02 eV)] might be attributed with lattice defects (oxygen vacancies, interstitials, etc.) which acts as unsaturated or dangling bonds in ionic oxides. Due to differences in ionic radii of Fe^{3+}, Li^+, and Ni^{2+} ions, the strain is induced in the spinel $Li_{0.3}Ni_{0.4}Fe_{2.3}O_4$ lattice [59]. This lattice strain might to create defect vacancies (XPS analysis). Voltage-current polarization characteristics of LNFO based HEC are obtained in **Figure 9(b)** [30]. Value of maximum offload output power is 13.77 mW, maximum offload current 15.3 mA and open cell voltage 0.9 V measured. The highly electronegative Ni^{2+} and

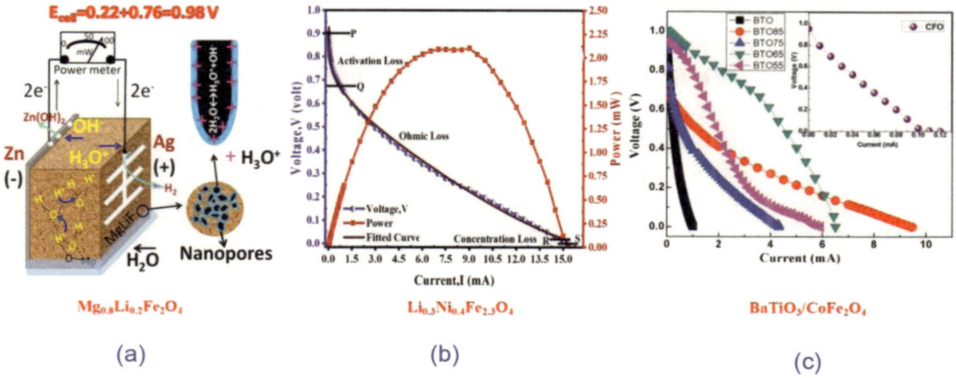

Figure 9.
(a) HEC working (schematic) due to water molecules dissociation and conduction of H_3O^+ and OH^- ions (zoom picture shows nanopores on pellet surface and further proton hopping inside nanopore to generates enough electric field) [12]. Voltage-current responses in (b) $Li_{0.3}Ni_{0.4}Fe_{2.3}O_4$ [30], (c) $(1-x)$ $BaTiO_3 - xCoFe_2O_4$ [7], HECs under soak water condition.

Fe^{3+} cations on LNFO surface attract a lone pair of electrons of oxygen present in water molecules. This strong attraction at octahedrally unsaturated surface cations leads to chemidissociation of water molecules into H_3O^+ and OH^-. At low current density, activation loss PQ is the potential required to overcome the energy barrier of electrochemical reaction. Ohmic loss QR is the resistance faced by ions during migration through the porous network. Concentration polarization loss RS signifies that the insufficient ions available at electrodes in the hyper-reactive state. The values of impedances of LNFO HEC in both dry and wet condition are given in **Table 3**.

3.7 Progress on $BaTiO_3$-$CoFe_2O_4$ multiferroic HEC

Lattice strain formation in multiferroic nanocomposites is highly induced defects associated vacancies [60]. Two constituent phases ferroelectric and ferromagnetic of multiferroic causes accumulation of charge carriers at grain boundaries and acts as active center for water molecule adsorption. **Figure 9(c)** shows the V-I results of multiferroic $(1-x)$ $BaTiO_3 - xCoFe_2O_4$ [x = 0.0 (BTO), 1.0 (CFO) 0.85 (BTO85), 0.75 (BTO75), 0.65 (BTO65), 0.55 (BTO55)] HEC synthesized by sol-gel process [7]. XRD pattern revealed the tetragonal BTO and spinel cubic CFO phases. For pure BTO, BTO85, BTO75, BTO65, and BTO55, the values of lattice constant for BTO are a = 0.399 nm, c = 0.403 nm; a = 0.399 nm, c = 0.401 nm; a = 0.400 nm, c = 0.402 nm; a = 0.400 nm, c = 0.403 nm; and a = 0.400 nm, c = 0.404 nm, respectively. Value of lattice constant for CFO phase a = 0.839 nm, 0.837 nm, 0.837 nm, 0.838 nm and 0.839 nm for pure CFO, BTO85, BTO75, BTO65, and BTO55 nanocomposite, respectively. Using method of XRD density, experimental (apparent) density, apparent density using Archimedes principle, the value of porosity is 2.304, 7.59, 10.8, 7.15, 6.18, and 5.54%, respectively measured for BTO, CFO, BTO85, BTO75, BTO65, and BTO55. From FESEM pattern, the average particle size, D = 80, 66, 70, 83, 277 and 300 nm, measured for BTO, CFO, BTO85, BTO75, BTO65, and BTO55, respectively. Maximum defects/vacancies in BTO85 nanocomposites are formed as confirm with PL emission. In **Figure 9(c)**, the pure BTO and CFO based HEC generated low current while nanocomposite cells exhibited current in mA. The observed variation in short circuit current, emf and output cell power is explained due to capacity of water dissociation that depends upon oxygen vacancies concentration, surface unsaturated bonds and nanoporosity. Oxygen vacancies provide dangling bonds in cell surface to attract more polar water molecules due to chemidissociation process by OH^- and H^+ ions. Pure BTO and CFO HECs generated short circuit current 2 and 1.2 mA, and emf 0.97 and 0.95 V, respectively. The BTO85, BTO75, BTO65, and BTO55 nanocomposite HEC generated short circuit current and emf 9.4 mA, 0.7 V; 4.2 mA, 0.90 V; 6.4 mA, 1 V; and 6.1 mA, 0.94 V, respectively.

3.8 Usability of iron oxide as HEC for future directions

Two major needs for good electric generation process in HEC of any metal-oxide material is that the nanomaterial to be highly defect states such as oxygen vacancies and porous formation or high porosity in pellet specimen. Iron oxides act as candidate material for HEC as discussed in the present chapter. Actually, iron oxides such as hematite, magnetite and maghemite are good source for Fe from +2 and + 3 valence states formation within oxygen vacancies. Therefore, the formation of oxygen vacancies in iron oxide based HEC depends upon the concentrations of Fe^{2+} and Fe^{3+} ions. The doping from transition metal and rare earth ions into hematite, magnetite, and maghemite nanomaterials, the surface oxygen vacancies might be increased. The concept of surface oxygen vacancies in iron oxides is already discussed in the introduction part. Non-stoichiometric nature of iron oxides also plays an important role for their usability in HEC. In case, the occurring of oxygen vacancies and valence fluctuations of Fe^{3+} and Fe^{2+} ions at magnetite surface attract water molecules towards its surface and chemidissociates it into H^+ and OH^- ions. Surface of Fe_3O_4 allows chemidissociation of water molecules due to attraction between the octahedrally coordinated unsaturated Fe^{2+} and Fe^{3+} cations and lone pair electron of oxygen in the H_2O molecule. Heterolytic splitting of the H_2O molecule takes place on Fe_3O_4 surface because relatively stable bondage of the highly electronegative Fe^{3+} and Fe^{2+} ions with lone pair electron of oxygen in H_2O. After heterolytic splitting, a Fe-OH bond is formed which lets the H^+ ion binds the neighboring oxygen atoms present on the surface to create another -OH surface group. Electrons trapped in oxygen vacancies act as dangling/unsaturated bonds attract the polar water molecule and unsaturated ferrous and ferric cations present on the surface of Fe_3O_4 lattice pull out OH^- ion from H_2O molecule and chemidissociates water into H_3O^+ and OH^- ions. Trapped H_3O^+ ions into nanopore develop a very high electric potential inside the nanopores due to Grotthuss chain reaction [10]. Spin density of Fe_3O_4 is determined to be 8.37×10^{24} spins/g, which is a measure of unsaturated/dangling (unpaired electrons) bonds present in the composition [8]. Water molecule interaction with hematite is highly dependent on the coordination number of Fe^{3+} ions along with surface composition, while dissociative adsorption of water is prominent on defective surface with coexisting Fe and O ions [9]. Non-stoichiometry in maghemite lattice can be enhanced by creating oxygen vacancies with existing surface iron vacancies increases surface reactivity significantly. In another case, surface of BTO-CFO nanocomposite unsaturated ions Fe^{2+}, Ti^{+3}, and oxygen vacancies attract polar water molecules to its closest approach followed by electron transfer and dissociated into hydronium and hydroxide ions [7].

4. Conclusions

The electricity generation by iron oxide based hydroelectric cell is a non-polluting and facile technique for green energy devices. Iron oxide nanoparticles in the form of hematite, magnetite, maghemite, ferrites, and ferrite multiferroic composite are prepared by solid state reaction, co-precipitation, and sol-gel methods. Hematite HEC delivered 30 mA current with an emf 0.92 V in wet deionized water condition. The maximum off-load power generated by magnetite and hematite is much higher than maghemite and ferrites based HEC. This is due to the enhancement capability of water molecule dissociation at surface Fe sites by highest number of oxygen vacancies formation to fallow surface hydroxylation process. The HEC of 4.08 cm^2 size of Li-doped Fe_3O_4 delivers I_{sc}, emf and P_{out} = 44.91 mA, 0.68 V, 30.80 mW, respectively. A 4.84 cm^2 size maghemite HEC generated stable and repetitive 19 mA I_{sc} with emf 0.85 V in deionized water. $Mg_{0.8}Li_{0.2}Fe_2O_4$ HEC of sizes 4.8 and 17 cm^2 in deionized water generated 8 and 82 mA I_{sc}, respectively. The $Li_{0.3}Ni_{0.4}Fe_{2.3}O_4$ HEC delivered output current density of 3.8 mA cm^{-2} which is two times higher with $Mg_{0.8}Li_{0.2}Fe_2O_4$ HEC (1.7 mA/cm^2).

Acknowledgements

The author K.C. Verma thankfully acknowledges the financial support by UGC of Dr. DS Kothari Post Doctorate Fellowship [No. F4-2/2006(BSR)/PH/16-17/0066] and CSIR-HRDG for SRA (Pool Scientist) fellowship Grant No. B-12287 [SRA (Pool No): 9048-A].

Author details

Kuldeep Chand Verma[1,2]* and Navdeep Goyal[2]

1 Materials Science and Sensor Applications (MSSA), CSIR-Central Scientific Instruments Organisation, Chandigarh, India

2 Department of Physics, Panjab University, Chandigarh, India

*Address all correspondence to: dkuldeep.physics@gmail.com; kcv0309@gmail.com

References

[1] Owusu PA, Asumadu-Sarkodie S. A review of renewable energy sources, sustainability issues and climate change mitigation. Cogent Engineering. 2016;**3**: 1167990

[2] Olabi AG, Abdelkareem MA, Wilberforce T, Sayed ET. Application of graphene in energy storage device—A review. Renewable and Sustainable Energy Reviews. 2021;**135**:110026

[3] Renewables 2021 Global Status Report. Paris: REN21 Secretariat; 2021. ISBN: 978-3-948393-03-8

[4] U.S. Energy Information Administration. International Energy Statistics. 2012. Available from: http://www.eia.gov/cfapps/ipdbproject/IEDIndex3.cfm?tid=2&pid=2&aid=2 [Retrieved: 18 October 2015]

[5] Gielen D, Boshell F, Saygin D, Bazilian MD, Wagner N, Gorini R. The role of renewable energy in the global energy transformation. Energy Strategy Reviews. 2019;**24**:38-50

[6] Sahoo SK. Renewable and sustainable energy reviews solar photovoltaic energy progress in India: A review. Renewable & Sustainable Energy Reviews. 2016;**59**: 927-939

[7] Shah J, Verma KC, Agarwal A, Kotnala RK. Novel application of multiferroic compound for green electricity generation fabricated as hydroelectric cell. Materials Chemistry and Physics. 2020;**239**:122068

[8] Jain S, Shah J, Dhakate SR, Gupta G, Sharma C, Kotnala RK. Environment-friendly mesoporous magnetite nanoparticles-based hydroelectric cell. Journal of Physical Chemistry C. 2018;**122**:5908-5916

[9] Jain S, Shah J, Negi NS, Sharma C, Kotnala RK. Significance of interface barrier at electrode of hematite hydroelectric cell for generating ecopower by water splitting. International Journal of Energy Research. 2019;**43**:4743-4755

[10] Gaur A, Kumar P, Kumar A, Shah J, Kotnala RK. An efficient green energy production by Li-doped Fe_3O_4 hydroelectric cell. Renewable Energy. 2020;**162**:1952-1957

[11] Shah J, Jain S, Gahtori B, Sharma C, Kotnala RK. Water splitting on the mesoporous surface and oxygen vacancies of iron oxide generates electricity by hydroelectric cell. Materials Chemistry and Physics. 2021;**258**:123981

[12] Kotnala RK, Shah J. Green hydroelectrical energy source based on water dissociation by nanoporous ferrite. International Journal of Energy Research. 2016;**40**(12):1652-1661

[13] Kotnala RK, Gupta R, Shukla A, Jain S, Gaur A, Shah J. Metal oxide based hydroelectric cell for electricity generation by water molecule dissociation without electrolyte/acid. Journal of Physical Chemistry C. 2018;**122**(33):18841-18849

[14] Kotnala RK, Shah J, Singh B, Kishan H, Singh S, Dhawan SK, et al. Humidity response of Li-substituted magnesium ferrite. Sensors and Actuators B: Chemical. 2008;**129**(2):909-914

[15] Verma KC, Goyal N, Singh M, Singh M, Kotnala RK. Hematite α-Fe_2O_3 induced magnetic and electrical behavior of $NiFe_2O_4$ and $CoFe_2O_4$ ferrite nanoparticles. Respiration Physiology. 2019;**13**:102212

[16] Verma KC, Sharma A, Goyal N, Kotnala RK. Ferromagnetism in

multiferroic $BaTiO_3$, spinel MFe_2O_4 (M = Mn, Co, Ni, Zn) ferrite and DMS ZnO nanostructures. In: Han M-G, editor. Electromagnetic Materials and Devices. Chapter 5. IntechOpen; 2019. pp. 1-24. DOI: 10.5772/intechopen. 82437

[17] Verma KC, Bhatt SS, Ram M, Negi NS, Kotnala RK. Multiferroic and relaxor properties of $Pb_{0.7}Sr_{0.3}[(Fe_{2/3}Ce_{1/3})_{0.012}Ti_{0.988}]O_3$ and $Pb_{0.7}Sr_{0.3}[(Fe_{2/3}La_{1/3})_{0.012}Ti_{0.988}]O_3$ nanoparticles. Materials Chemistry and Physics. 2010;**124**:1188-1192

[18] Teja AS, Koh PY. Synthesis, properties, and applications of magnetic iron oxide nanoparticles. Progress in Crystal Growth and Characterization of Materials. 2009;**55**:22-45

[19] Wu W, Wu Z, Yu T, Jiang C, Kim WS. Recent progress on magnetic iron oxide nanoparticles: Synthesis, surface functional strategies and biomedical applications. Science and Technology of Advanced Materials. 2015;**16**:023501

[20] Mathpal MC et al. Ferrites as an alternative source of renewable energy for hydroelectric cell. In: Sharma SK, editor. Spinel Nanoferrites. Topics in Mining, Metallurgy and Materials Engineering. Cham: Springer; 2021. DOI: 10.1007/978-3-030-79960-1_13

[21] Cornell RM, Schwertmann U. The Iron Oxides: Structures, Properties, Reactions, Occurrences and Uses. Weinheim: Wiley; 2003

[22] Zysler RD, Fiorani D, Testa AM. Investigation of magnetic properties of interacting Fe_2O_3 nanoparticles. Journal of Magnetism and Magnetic Materials. 2001;**224**(1):5-11

[23] Margulies DT, Parker FT, Rudee ML, Spada FE, Chapman JN, Aitchison PR, et al. Origin of the anomalous magnetic behavior in single crystal Fe_3O_4 films. Physical Review Letters. 1997;**79**(25):5162-5165

[24] Zhu W, Winterstein JP, Maimon II, Yin Q, Yuan L, Kolmogorov AN, et al. Atomic structural evolution during the reduction of α-Fe_2O_3 nanowires. Journal of Physical Chemistry C. 2016;**120**(27):14854-14862

[25] Spreitzer D, Schenk J. Reduction of iron oxides with hydrogen—A review. Steel Research International. 2019;**90**:1900108

[26] Setvin M, Wagner M, Schmid M, Parkinson GS, Diebold U. Surface point defects on bulk oxides: Atomically-resolved scanning probe microscopy. Chemical Society Reviews. 2017;**46**: 1772-1784

[27] Singh D, Verma KC. Magnetic properties of Heusler alloys and nanoferrites. In: Sahu DR, editor. Magnetic Skyrmions. IntechOpen; 2021. DOI: 10.5772/intechopen.95466

[28] Ribeiro FN, Pedroza LS, Souza FL, Dalpian GM. Surface Fe vacancy defects on haematite and their role in the light-induced water splitting in artificial photosynthesis. Physical Chemistry Chemical Physics. 2017;**19**:31410-31417

[29] Ovcharenko R, Voloshina E, Sauer J. Water adsorption and O-defect formation on Fe_2O_3 (0001) surfaces. Physical Chemistry Chemical Physics. 2016;**18**:25560-25568

[30] Saini S, Shah J, Kotnala RK, Yadav KL. Nickel substituted oxygen deficient nanoporous lithium ferrite based green energy device hydroelectric cell. Journal of Alloys and Compounds. 2020;**827**:154334

[31] Kharbanda P, Madaan T, Sharma I, Vashishtha S, Kumar P, Chauhan A,

et al. Ferrites: Magnetic materials as an alternate source of green electrical energy. Heliyon. 2019;**5**:e01151

[32] Chernyshova IV, Hochella MF Jr, Madden AS. Size-dependent structural transformations of hematite nanoparticles. Physical Chemistry Chemical Physics. 2007;**9**(14):1736-1750

[33] Singh S, Singh M, Ralhan NK, Kotnala RK, Verma KC. Ferromagnetic and dielectric properties of Ni Zn Fe O nanoparticles prepared via chemical combustion route. Advanced Science, Engineering and Medicine. 2014;**6**(6):688-693

[34] Singh S, Singh M, Ralhan NK, Kotnala RK, Verma KC. Improvement in ferromagnetism of $NiFe_2O_4$ nanoparticles with Zn doping. Advanced Materials Letters. 2012;**3**(6):504-506

[35] Ding Y, Kong XY, Wang ZL. Interface and defect structures of Zn-ZnO core-shell heteronanobelts. Journal of Applied Physics. 2004;**95**(1):306-310

[36] Lehtinen O, Tsai IL, Jalil R, Nair RR, Keinonen J, Kaiser U, et al. Non-invasive transmission electron microscopy of vacancy defects in graphene produced by ion irradiation. Nanoscale. 2014;**6**(12):6569-6576

[37] Yin S, Ellis DE. H_2O adsorption and dissociation on defective hematite (0001) surfaces: A DFT study. Surface Science. 2008;**602**(12):2047-2054

[38] Jones F, Rohl AL, Farrow JB, Bronswijk W. Molecular modelling of water adsorption on hematite. Physical Chemistry Chemical Physics. 2000;**2**(14):3209-3216

[39] Joseph Y, Kuhrs C, Ranke W, Ritter M, Weiss W. Adsorption of water on FeO (111) and Fe_3O_4 (111): Identification of active sites for dissociation. Chemical Physics Letters. 1999;**314**(3-4):195-202

[40] Yin S, Ma X, Ellis D. Initial stages of H_2O adsorption and hydroxylation of Fe-terminated α-Fe_2O_3 (0001) surface. Surface Science. 2007;**601**(12):2426-2437

[41] Stern M, Geaby AL. Electrochemical polarization I. A theoretical analysis of the shape of polarization curves. Journal of the Electrochemical Society. 1957;**104**(1):56-63

[42] Ihonen J, Jaouen F, Lindbergh G, Lundblad A, Sundholm G. Investigation of mass-transport limitations in the solid polymer fuel cell cathode. Journal of the Electrochemical Society. 2002;**149**:448-454

[43] Andre D, Meiler M, Steiner K, Wimmer C, Soczka-Guth T, Sauer DU. Characterization of high-power lithium-ion batteries by electrochemical impedance spectroscopy. I. Experimental investigation. Journal of Power Sources. 2011;**196**(12):5349-5356

[44] Joseph Y, Ranke W, Weiss W. Water on FeO (111) and Fe_3O_4 (111): Adsorption behavior on different surface terminations. The Journal of Physical Chemistry B. 2000;**104**:3224-3236

[45] Zhang X, Niu Y, Meng X, Li Y, Zhao J. Structural evolution and characteristics of the phase transformations between α-Fe_2O_3, Fe_3O_4 and γ-Fe_2O_3 nanoparticles under reducing and oxidizing atmospheres. CrystEngComm. 2013;**15**:8166-8172

[46] Lei W, Liue Y, Si X, Xu J, Du W, Yang J, et al. Synthesis and magnetic properties of octahedral Fe_3O_4 via a one-pot hydrothermal route. Physics Letters A. 2017;**381**:314-318

[47] Verma KC, Singh D, Kumar S, Kotnala RK. Multiferroic effects in $MFe_2O_4/BaTiO_3$ (M = Mn, Co, Ni, Zn) nanocomposites. Journal of Alloys and Compounds. 2017;**709**:344-355

[48] Chan SH, Khor KA, Xia ZT. A complete polarization model of a solid oxide fuel cell and its sensitivity to the change of cell component thickness. Journal of Power Sources. 2001;**93**:130-140

[49] Jo'm B, Reddy AK. Modern Electrochemistry: An Introduction to an Interdisciplinary Area. 1970th edition. Plenum Press, Springer; 1973

[50] Jaouen F, Lindbergh G, Sundholm G. Model based PEM fuel cell state-of-health monitoring via ac impedance measurements. Journal of the Electrochemical Society. 2002;**149**(4):A437-A447

[51] Verma KC. Synthesis and characterization of multiferroic $BiFeO_3$ for data storage. In: Luo Y, Wen J, Zhang J, editors. Bismuth—Fundamentals and Optoelectronic Applications. IntechOpen; 2020. DOI: 10.5772/intechopen.94049

[52] Barrn V, Torrent J. Evidence for a simple pathway to maghemite in Earth and Mars soils. Geochimica et Cosmochimica Acta. 2002;**66**:2801-2806

[53] Lovely GR, Brown AP, Brydson R, Kirkland AI, Meyer RR, Chang L, et al. Observation of octahedral cation coordination on the (111) surface of iron oxides nanoparticles. Applied Physics Letters. 2006;**88**:093124

[54] Verma KC. Diluted magnetic semiconductor ZnO: Magnetic ordering with transition metal and rare earth ions. In: Sahu DR, Stavrou VN, editors. Magnetic Materials and Magnetic Levitation. Chapter 6. IntechOpen; 2020. pp. 1-23. DOI: 10.5772/intechopen.90369

[55] Sadat ME, Baghbador MK, Dunn AW, Wagner HP, Ewing RC, Zhang J, et al. Photoluminescence and photothermal effect of Fe_3O_4 nanoparticles for medical imaging and therapy. Applied Physics Letters. 2014;**105**:091903

[56] Onsten A, Weissenrieder J, Stoltz D, Yu S, Gothelid M, Karlsson UO. Role of defects in surface chemistry on Cu_2O (111). Journal of Physical Chemistry C. 2013;**117**:19357-19364

[57] Mazen SA, Abdallah MH, Nakhla RI, Zaki HM, Metawe F. X-ray analysis and IR absorption spectra of Li-Ge ferrite. Materials Chemistry and Physics. 1993;**34**:35-40

[58] Aravind G, Raghasudha M, Ravinder D. Electrical transport properties of nano crystalline Li-Ni ferrites. Journal of Materials. 2015;**1**:348-356

[59] Ramesh M, Rao GSN, Samatha K, Rao BP. Cation distribution of Ni-Cu substituted Li-ferrites. Ceramics International. 2015;**41**:1765-1770

[60] Verma KC, Singh M. Processing techniques with heating conditions for multiferroic systems of $BiFeO_3$, $BaTiO_3$, $PbTiO_3$, $CaTiO_3$ thin films. In: Qin G, editor. Thermoelectricity-Recent Advances, New Perspectives and Applications. IntechOpen; 2021. DOI: 10.5772/intechopen.101122